TILTWALLISM

TILTWALLISM

A TREATISE ON THE ARCHITECTURAL POTENTIAL OF TILT WALL CONSTRUCTION

JEFFREY BLAINE BROWN

Published in Australia in 2014 by
The Images Publishing Group Pty Ltd
ABN 89 059 734 431
6 Bastow Place, Mulgrave, Victoria 3170, Australia
Tel: +61 3 9561 5544 Fax: +61 3 9561 4860
books@imagespublishing.com
www.imagespublishing.com

National Library of Australia Cataloguing-in-Publication entry

Title:	TILTWALLISM : a treatise on the architectural potential of tiltwall construction / Jeffrey Brown
Author:	Brown, Jeffrey, author
ISBN:	9781864705751 (hardback)
Subjects:	Concrete construction Building technique Architecture

Dewey Number: 693.5

Edited by Sabita Naheswaran

Designed by Powers Brown Architects

Pre-publishing services by United Graphic Pte Ltd, Singapore

Printed by Everbest Printing Co. Ltd., in Hong Kong/China on 140gsm
GoldEast Matt Art

IMAGES has included on its website a page for special notices in
relation to this and our other publications.
Please visit www.imagespublishing.com

For Joe Powers, AIA.

This was always your tune,

I just wrote some lyrics for you.

TABLE OF CONTENTS

INTRODUCTION

Architectural education and its relationship to architectural practice have changed a great deal across the last 50 years. Design theories have come and gone, and while the design studio remains the international pedagogical anchor for most architectural curricula, a direct connection between the design studio and architectural practice, perhaps even between the studio and *making architecture*, has waned. Architectural education has become a discipline unto itself, apart from the practice of architecture.

Several factors have fueled this shift. Guidelines for faculty members' advancement have often been standardized across the many disciplines represented within today's large public universities. For evidence of achievement in academia, tenure and promotion guidelines now often favor more traditional forms of faculty-member scholarships compared to more discipline-specific activities such as writing poetry, making art, or producing architecture. Because of the ever-present, heightened contemporary threat of litigation, university committees and administrators seek university-wide objectivity with clear, irrefutable consistency in evaluating faculty members' scholarships.

Meanwhile more "peer reviewed" venues for presenting and publishing papers (such as the conferences organized by the Association of Collegiate Schools of Architecture) have developed and grown. Thus, instead of producing buildings, today more and more architects teaching in architecture schools write papers on focused research topics to satisfy the "publish or perish" requirements for advancement in academia. Simultaneously, practitioners increasingly view architects in academia and their scholarship as being products of ivory towers and as being irrelevant to the daily challenges encountered in architectural practice.

This book is written by an anomaly to this trend. Here is an architectural teacher who runs a practice and makes architecture, and who is perhaps even more unusual because he is simultaneously exceptionally well versed in architectural history and theory, even among academics. Perhaps rarer still, Jeffrey Brown also sees no implicit conflict between the goal of making architecture imbued with our highest goals and our concurrent adherence to the constraints produced by the realities of typical American commercial enterprise!

When I think of where we are now regarding the relationship between architectural education and architectural practice and how we got here, I find this a very hopeful book. Jeffrey senses a current trend of a heightened interest in re-engaging the actual materiality and processes of making buildings. He also sees the possibility of new architectural theory arising at the cusp of this renewed interest and our making architectural form, and he proposes that theory should be produced from within the disciplines of design, architecture, and engineering through questions arising from actions, and processes related to making buildings. *TILTWALLISM* encourages teacher and practitioner alike to look inside architecture for meaningful research and theory topics rather than (or at least in addition to) outside architecture. Here design theory could spring from designing and making buildings rather than being imported and overlaid from "outside" architecture. Jeffrey also sees the possibility of architectural theory being a continuum, with testing. Unlike some more esoteric forms of architectural theory popular in the last century, *TILTWALLISM* explores the potential for architectural theory to be proven through its application (as in other disciplines), but importantly, what is "tested" in this instance is a synthetic, whole building design rather than a small isolated piece of a building as is often the case in much academic research today.

TILTWALLISM champions Tilt Wall construction as a particularly rich territory for this research and for formal invention. In Jeffrey's view Tilt Wall is a new form of "global vernacular" that enables a response to the ubiquitous need for cheap, direct, and fast construction. But *TILTWALLISM* also sees the architectural value of Tilt Wall as not implicitly constraining its efficiency as a construction process or its economy, and conversely, the economy and efficiency of Tilt Wall construction is not seen here as diminishing its architectural potential. Because it operates at architecture's base level of construction and structure, the system is seen as capable of fostering clarification and refinement within architectural proposals. Here Tilt Wall, as a way and means of making, is viewed as primarily structure; and when structure (here the Tilt Wall system) is coupled to program and design skill, elegance is possible. Historically a similar formulation for elegance can be found in many works—particularly in the direct and efficient structure and beautiful design elegance of Robert Maillart's work. *TILTWALLISM* shows that with Tilt Wall, function, construction, and meaning are inextricably linked. I am reminded here of a quote from the Pritzker Prize and RIBA Royal Gold Medal winning architect Rafael Moneo, who spoke of architecture "arriving" at the point "when our thoughts about it acquire the real condition that only materials provide".

Jeffrey also offers an egalitarian, democratic stance for access to design excellence. In *TILTWALLISM* minimal budgets and real economic metrics do not eliminate the possibility of making architecture, and mainstream architectural practice can also be critical, experimental or even avant-garde. This is admittedly and demonstrably rare—but here Tilt Wall is offered as an economical and accessible way to engage the experiment. Jeffrey is attempting to "raise consciousness", and he maintains that meaningful architectural operations are possible within capitalistic society and should be accessible to all, and to all projects regardless of budget. He specifically champions the use of Tilt Wall applications in low-budget projects for small business operations, and he insists that these buildings should not be excluded from architecture. The work of his firm shows us how this can be done.

TILTWALLISM encourages us, as architects, to move from seeking clients toward the broader and more creative goal of inventing markets. This path is demonstrated in the successful and award-winning work of Powers Brown Architecture. Also included here is an illuminating historical account of Tilt Wall—summarized as a process first embraced by architects, only to later become a mainstream technology largely ignored by architects. Ultimately Jeffrey shows us through his work and through the work of his firm that it is possible to directly engage and accept commercial economic limits while conceding nothing in terms of architectural goals and potential.

Joe Mashburn FAIA
Professor Emeritus and Former Dean,
Gerald D. Hines College of Architecture
at the University of Houston

ACKNOWLEDGEMENTS

All architects have had a formative experience or two. While there are, as I assume there are for many of my colleagues, too many to mention at any single polite setting, two of mine seem to contribute to two organizing themes in this book: that of the theory of the everyday, and the role of the market loosely defined. For the former I recall my first encounter with the writing of Robert Venturi in Complexity and Contradiction in Architecture, and soon thereafter, his collaboration with Denise Scott Brown and Steven Izenour, producing Learning from Las Vegas. I was, as trained by my studio time, repelled at their architecture, but something about the idea of teasing latent content out of the everyday, and using it as THEORY (as I understood it back then) never left me. I have spent almost every day thereafter riffing on that framework for my own pursuits. The second is more abstract in both memory and effect. I vaguely recall a dinner, in 1985 or 1986, hosted by my then mentor Richard McBride and attended by Colin Rowe and Jaqueline Robertson at the Stoneleigh P Hotel in downtown Dallas. I say vaguely because I recall almost nothing in great detail, and understood, as I remember, very little—but for one big thing. A discussion erupted around the notion, posited by Colin, that good designers just didn't or perhaps couldn't make money, and this was disagreeable to Jaqueline for some reason. At some point the words "capitalism" and "market" were employed along with some snippets of Marx. This was my first inkling about the relationship between architecture and commodity; a thought that has woven itself into my own conceptions every day since.

As for acknowledging more direct influence, many others have helped me directly on this project. I first began to think about writing *TILTWALLISM* while teaching at the University of Houston's College of Architecture under Joe Mashburn FAIA. Dave Hickey laments his own introduction to academia by noting that all that is interesting as subject matter is free, prosaic and entrepreneurial outside of the academy, but gets owned, territorialized and mastered within it. (He prefers the outside). Joe knows about the stultifying effects of this condition and, during his time as Dean, managed to create fissures of opportunity, encouragement and dissent within its structure—encouraging the majority of us that were practicing teachers to "do our thing". No small accomplishment, and done without almost any acknowledgement. I hope others will take advantage of what happened then—like I did—by finishing what has been started. Others like Geoffrey Brune FAIA and Donna Kacmar FAIA have also been "accidental encouragers" of this author over time.

Closer to home I want to thank John Cadenhead, Daphne Dow, and James "Banksy" Gavin for many hours spent away from our practice focused on this project. John Cadenhead has been my right hand on so many projects over the years; his authorship of the graphic section far exceeded my vision in suggesting it. Thanks also to Desta Kimmel, Randal Hall PhD., and Mitch Bloomquist, for reading parts and wholes of various phases of the text. The participants in the case study section, namely Rand Elliot FAIA who broke the ice by agreeing to contribute, Steven Holl, Carlos Jiménez, Patkau Architects, Rob Quigley, and cunningham architects (sic) have my deepest appreciation; their work will lend some gravity and historical basis to both the book and subject.

There have been many colleagues with whom I have discoursed with over minor questions and major concepts, including those who contributed by rewriting, critiquing and in some cases flat out refuting chapters pertaining to their expertise: Jeff Griffin PhD., Kevin Rogge, Mark Reed, Steve Salverino, Kent Penrod, Paul Coonrod, Carleton Riser, David Tomasula, Don Greive Eng., Brad Bihner Eng., and Adam Cryer Eng.,—thanks to all of you, and to those of you whom I am sure I missed.

Finding a publisher who is as open-minded and enthusiastic as Paul Latham at IMAGES is a long shot. But then again in today's environment getting any publisher interested in a subject that has not been established in the market is a long shot. It seems like publishers would be looking for the uncovered and undiscovered, but when you look on the shelves at bookstores and see 11 books on prefab houses and Frank Lloyd Wright, you know that the market dictates what will sell and what won't. But somebody wrote the first prefab book and somebody took a long shot publishing it. Paul Latham is my long-shot guy, and I thank him for it.

Jeffrey Blaine Brown, Houston, 2014.

PROLOGUE

"There was not, in the last quarter of the nineteenth century, much agreement among theoreticians and architects as to what constituted rational building, as to how structure was related to form, or to as to how construction affected style ... Even the most iconoclastic of the modernists were seldom able to discuss construction without some knowledge of and some reference to precedent ... How could the same architects be so perceptive of one phenomenon and so blind to another?"

Edward Ford, **Details of Modern Construction**

The broad context
of the **tilt wall problem**

In considering the problem of a building method such as Tilt Wall construction as a primary source of formal invention, a bit of context is in order. In the last twenty years or so debate has arisen around just such ideas and from multiple sources and motivations. Kenneth Frampton has been at the forefront of thinking about the relationship between tectonics and institutionalized architectural form. A convenient and compact case in his own words deals tectonics, the bandwidth of making Tilt Wall appears to fall within, something less than the upper hand:

> ... one might say that the autonomy of architecture is determined by three inter-related vectors; typology (the institution), topography (the context) and tectonics (the mode of construction). It should be noted that neither the typological nor the tectonic are neutral choices in this regard and that what can be achieved with one format and expression can hardly be realized by another.

> On balance, the formal parti is of greater import that the tectonic, for obviously the selection of the type as the basic spatial order has a decisive impact on the result, however much the constructional syntax may be elaborated in the course of development. The primacy of the type perhaps makes itself most evident in the basic relationship between building and architecture: for where building tends to be organic, asymmetrical, and agglutinative, architecture tends to be orthogonal, symmetrical, and complete. These distinctions would not be so crucial were it not for the fact that building and architecture tend to favor the accommodation of different kinds of institutional form.[i]

This suppression of the potential primacy of tectonics in the generation of meaningful form was counterbalanced, but importantly not necessarily counter-posed by arguments for a broader interpretation of what is considered to be architectural. These formulations dealt in the main with social aspects of architecture's generative purpose—asking questions such as why so little is considered to be circumscribed by architecture, or why it is so much (80 percent) of all that is built does not involve architects and thus is excluded from institutionalized architectural criticism, discourse and alarmingly practice. These cases argued for a broader circle of considerations such as social, political and economic forces to be considered in form generation as well as in its interpretation and evaluation.[ii] Here, tectonics like Tilt Wall could become a major component of meaning. This overlapping condition, on the one hand tectonics as an important but secondary criterion and on the other a possible primary consideration, approximately describes the current environment of tectonic consideration—tectonics is either institutional or instrumental in many current works, it provides the expressed poetry of Ando's forms or the basis of the technological abstraction of many of Foster's. In both cases teasing open the closed problems and limited propositions that are considered capital "A" architecture are the goal. Very recently contemporary historic fluctuations have reopened new space for debating the role tectonics will play in architectural form making, and at the very least invite a deeper exploration of its formal systems such as Tilt Wall construction. At best these happenings (like the Great Recession) encourage the continued broadening of architectural considerations away from the closed and problematically limited offerings that seem to dominate discourses.

There is frequent recent speculation in a wide range of periodicals, online webcasts and professional industry specific economic outlooks that the Architecture / Engineering / Construction (AEC) industry may not return to pre-boom levels of work for perhaps a decade. Thus there are fewer projects to design and less money available to build them. Articles in *Architect* and *Architectural Record* have posed the question: is the era of trophy building over? This has fueled renewed interest and pressure in exploring economical methods of financing, constructing and conceiving all building types. If one is to believe the predictions, this is a challenge that may face the allied professions in the building industry for some time to come. This dilemma sponsors an obvious question for design professionals: will design innovation suffer at the proposition of less costly methods of building?

Paradoxically, perhaps not. According to the Tilt-Up Concrete Association (TCA), Tilt Wall construction is one of the fastest growing methods of construction worldwide. In its simplest form, Tilt Wall construction is

a technique in which the exterior walls of almost any building type are cast onsite, indeed on the buildings slab most often, and then "tilted" into place with a crane. It can accommodate scales as small as residential and as large as one-million-square-foot buildings and has no geographic or topographic limitations. Given the recent experience of and future projections regarding the prolonged downward pressure on cost of construction, Tilt Wall construction is poised to take its place alongside the digital revolution and sustainability as the way forward for many projects. Tilt Wall construction offers both the design flexibility and the economic methodology to address the dilemma posed by downward cost pressure and a desire for maximum design flexibility. Because it is site cast, and uses existing aspects of the permanent structure, it has cost advantages in terms of time, material and labor.

It is spreading into new building types in the States and is employed and becoming prevalent in Canada, Australia, South America, Russia and the Dominican Republic, to name a few. There are efforts under way in Haiti to train locals to use the Tilt Wall methodology to create safe,[iii] fast and yet reasonably disaster resistant indigenous housing. The Middle East, with its economy projected at 10 percent growth even during the current recession, has several major design and construction organizations exploring this way of building. While the subject has benefitted from articles and projects published across trade organization websites and blogs, it has received no formal treatment in the academic media or professional architectural press. Curiously, a handful of well-known architects have started to dabble in the method on individual projects. Steven Holl, David Chipperfield, Mack Scoggin, Rob Quigley, Rand Elliot and others have achieved projects which have garnered the highest level of recognition with the American Institute of Architects awards programs and coverage in the national press in addition to inclusion in monographs and regional publications. Yet this alone has not proved sufficient to ignite interest in the great majority of mainstream practices that could, in the given context, benefit from the potential of its formal possibilities and budgetary advantages. One substantive reason for this is that Tilt Wall has received no formal treatment in architectural literature as of yet. This book is designed to remedy this.

Why tilt wall as a subject?

Tilt Wall has the potential to re-engage architecture with everyday building types and site specific form-making that have been abandoned and neglected by architects on many fronts. This book is offered as a provocation for them to recognize Tilt Wall construction as a rich source of design potential. Here, it is not to be questioned whether the study of novel form is necessary, nor is it to be questioned whether the importation of ideas outside of architecture, related to process or to the interpretation and understanding of form, are of importance. Both of those assertions are granted. Rather it is to be proposed that there is unexploited territory for investigation and meaningful creative inspiration within the circumscription of the process of building itself that seems historically to suffer at the expense of these pursuits. While this book is informed by the discovery and use of Tilt Wall as a way of building architecture, it was inspired as somewhat of a reaction to the ecumenical acceptance of the notion of "transfer technology" as proposed by Kieran and Timberlake in their influential book *Refabricating Architecture*, and as a reaction to the trickle-down enthusiasm of Reiser's and Umemoto's *Atlas of Novel Tectonics*. In both cases the excitement stimulated by these works bridges from my students to the many derivative academic and professional exercises related to these seminal texts.

In the former case, the fascination with outside technologies and processes that might serve as models for the transformation of the production and design strategies in architecture, while prescient, relies upon an exterior focus that has structured architectural thinking and theory on various, perhaps even cyclical, occasions. While there is much to be admired in Kieran's and Timberlake's research, it is certainly not the first time we have looked to disciplines outside of architecture for means to understand and produce content from within it. Le Corbusier of course imagined architecture as the potential product of mass production inspired by engineering and manufacturing processes. Peter Eisenman famously looked to Chomsky to understand the notion of architecture as a language, thus creating much of what became institutionalized theory in the 80s and 90s. In all cases of "looking outside", there is a counter argument to be made along the lines that architecture need not be an analog to alternative disciplines in order

to generate meaningful production. Indeed we have been reminded in past arguments it is quite capable of doing so from within.[iv] The argument here will be an examination of Tilt Wall as a technology residing within the discipline of architecture, engineering and design. It does so in the same spirit of Kieran and Timberlake's search for the way forward in architecture, while remaining cognizant of the idea that we have not begun to exhaust the possibilities of the processes and technologies that circumscribe the discipline itself. The irony of the bankruptcy of the very automobile industry they have modeled much of their argument on is not lost in the proposition that it is within the discipline that we should look for existing but unexploited methods and opportunities.

Regarding the latter, Reiser and Umemoto in many ways embody the avant-garde fascination for the accommodation of the digital in architecture, certainly at the most considered and deliberate level. They are not perpetrators of absent-minded experimentation; indeed their argument comprises the most insightful analysis of both the potential and pitfalls of the digital revolution in form making. Yet as the avant-garde rushes to break new ground, incorporate new technologies and materials into the field of architecture and create novel forms, one could argue a paradoxical result; an ever-narrowing scope of projects and building types that are considered capable of embodying architectural potential. Architects have played a dangerous game, ceding territory to other disciplines—builders, contractors and agents of client interface—to the extent that it has undoubtedly transformed the profession in the span of only a few generations. As we did by abandoning housing as an area of serious practice at the beginning of the 20th century,[v] so we are doing now with office buildings, warehouses and factories.[vi] Shopping centers, manufacturing plants and, in far too many cases, schools are left behind as containers of architectural potential in favor of museums and a select few high budget houses. Part of this disinterest stems from arguments made by Margaret Crawford that in these building types, many of the key decisions have been made prior to the commission of an architect.[vii] Market forces channeled through brokers, "comparables" and investment norms preclude the invention, at a DNA-level, of everyday buildings. Thus architects are often left with the task of skinning carefully contrived economic diagrams, providing money making scaffolds with an image often to be built with low budget materials and construction. This becomes the rub; just what has the *Atlas of Novel Tectonics* to do with the majority of the built environment? How can the arguments in Reiser and Umemoto's catalog, relevant as they may be in academic dialog and the design of museums or houses for the now termed "one percent", apply to the market-driven blight-scape we experience on our drive home every evening? Are we experimenting on the future at the expense of the present?

Both recent tendencies betray architecture's fascination with technology. Technology is an original element of design thinking—relating meaningful form to the manner in which it can be constructed. At times however, technology as a method seems to have supplanted form making as the primary focus of the architect. From the pyramid builders, to Brunelleschi's dome, to the encyclopedic drawings during the Enlightenment and into the Industrial Revolution, technology was progressive and progressive architecture was entwined with it. But Heidegger warned that technology alone was not problematic with regards to the benefit it brings but rather its emergence as an autonomous force.[viii] For architecture this manifests itself in the dearth of major innovation into truly new ways of building everyday projects, and in vestigial methods developed along the way that never seemed to progress nor ever be proven obsolete. This is both the dilemma and source of possibility in Tilt Wall construction.

By illustrating the potential of approaching architectural exploration as a result of construction, particularly a "low technology" form of construction, this book is meant to stimulate an interest in Tilt Wall for architectural designers. Inarguably, Tilt Wall is fast becoming a mainstream method of construction across the globe, yet architecture remains strangely disengaged with investigating it. The most apparent reason is that, as a method of construction, it does not immediately or obviously fall into the category of architecture or its purview, at least at face value. Certainly there are arguments to be made against form, and thus architecture, driven by construction alone. Yet there is a kind of dirty reality to the effectiveness of Tilt Wall "technology" pushing ever further upstream, reaching the point of K-12 schools and universities, hospitals and other major institutions, once the exclusive realm of form-driven design. If David Hickey is correct in his proposition that it is "institutions" architecture must re-engage,[ix] then these developments ought to ignite an architectural interest in the low tech

and viral spread of Tilt Wall construction. The paradox is that architects, via technology, have seldom in recent times been more enmeshed in how things are made and what they are made from. Recently Alejandro Zaera-Polo has observed that architects desire to have their works read as an actualization of the potentials of building technology rather than, as the moderns may have, the optimal solution to technical concerns, perhaps even economic ones.[x]

This is precisely the dilemma of Tilt Wall construction as it awaits further integration into architectural legitimacy. It is a "low" form of building methodology utilized in commercial construction by "mainstream" architects and builders. This implies that there are "high" forms of construction reserved perhaps for culturally significant buildings and institutions. They are often complex hybrid structures that consume great portions of the budget in supporting novel form alone—often contributing very little to the overall content of the building as regards meaning. Ando's use of load-bearing cast concrete or Rem Koolhaas's CCTV building's use of structural cladding are examples. But the application of these innovations in everyday architecture remains remote. So, can some low forms of construction prove to be valid approaches to contributing content to architecture? What does it communicate to build this way? Can Tilt Wall be seen as an "actualization" versus a mere vehicle of economic expediency? Is there a difference in the end?

These questions remain unexplored as a result of the resistance by establishment academics and practitioners to Tilt Wall construction's inevitable advance into architecture's space—the subject is explored in the following pages. At a time of ever more intense research into methods for building novel form, one of the most prevalent techniques for building, which was invented over 100 years ago, has been totally overlooked by architecture.

Organization

The book is organized as two independent overlapping systems. The numbered chapters cover history, theory and interpretation of the method in mostly textual format. They are designed to read sequentially and form a more or less continuous argument that obtains objectivity (as it is possible). Dividing them is a kind of second book embedded between chapters titled "Excursus". They cover in chronological order the actual "how to" of the Tilt Wall method: an example of using it as a market-based form generator, and a section on case studies covering the far too few significant uses of the method by well-known architects, or on well-known projects. These sections are graphically driven, written in independent voices and can be interpreted almost independently from the text sections. They are unabashedly subjective.

A brief note on nomenclature

I have used the term "Tilt Wall" throughout this book as a preferred alternate to "Tilt Up". In the latter case, my feeling is that one could assume tilt up applies to anything, almost as a transitive verb. Tilt Wall is more specific in my mind as to just what it is being "tilted up". However in many parts of the country and indeed world they are interchangeable.

i Frampton, Kenneth. "Reflections on the Autonomy of Architecture: A Critique of Contemporary Production", *Out of Site: A Social Criticism of Architecture*. Bay Press, Seattle,1991.

ii There were numerous proponents of this approach including Diane Ghirardo.

iii Sherman Balch, Tilt Up Concrete Association member.

iv I am here referring to Jorge Silvetti's basic argument in the well-known essay "Beauty of Shadows". Silvetti makes a fundamental argument in service of autonomous architecture.

v The common argument is that architects are involved in something less than 10 percent of all residential construction. This is anecdotally substantiated by Margaret Crawford in *Out of Site: A Social Criticism of Architecture* (Bay Press, Seattle, 1991).

vi Crawford, M. "Can Architecture be Socially Responsible?", *Out of Site: A Social Criticism of Architecture*. Bay Press, Seattle,1991.

vii Ibid., pg 30.

viii Leach, N. "Introduction", *Rethinking Architecture: A Reader in Cultural Theory*, Routledge, New York, 1997.

ix Hickey, David. 'On Not Being Governed', *The New Architectural Pragmatism: Harvard Design Magazine Reader*, University of Minnesota Press, Minneapolis, 2007. Hickey, by institutions, means the formalization of standards and critiques that suppress innovation at the expense of invention and originality or creativity. He elaborates this theme elsewhere in his writing.

x Zaera- Polo, Alejandro. "A Scientific Autobiography", *Harvard Design Magazine*, Vol. 21, Fall / Winter 2004.

"Well, one thinks about the lawyer with a whole library bound in blue morocco behind him. This is the inventory of cases bearing upon the specific case that he is requires to judge. So simply to pronounce a legal innovation, to discriminate the new, our jurist is obliged to consult the old and the existing; and it is only by reference to these that genuine innovation can be proclaimed. For are not precedent and invention the opposite sides of the same coin? I think a better topic might have been: How does the new invade the old and how does the old invade the new?"

Colin Rowe, Letter to the Editor,
The Harvard Architecture Review, Vol. 5.

The history of Tilt Wall construction has to be teased out, like a beneficial genetic mutation, from the seam between the pre-modern formative period and that of late Industrial Revolution hubris in the early 20th century. Imbricated in construction history, Tilt Wall construction has concomitantly been retarded as an independent architecturally important development. Steel frame construction, the elevator and poured-in-place concrete methods, as didactic independent building techniques, have architectural influence on aesthetic design potential up to the current day. Tilt Wall construction, invented at the same time as these methods and techniques, has also persevered and, indeed in many markets, has become a globally dominant approach to building; yet it has never received intense investigation regarding its architectural design potential, let alone its unique developmental history. As the Tilt Wall method became dominant as a commercial and increasingly popular architectural method, its history is all the more important to understand. Many of its methodological siblings like those described above have been radically modified, totally integrated into synthesized hybrid approaches, or have become outdated, yet the proposition of Tilt Wall construction as a particularly potent architectural approach to conceiving of form has never been stronger.

Several conditions illustrate the conundrum as to why it has not been more celebrated even as it has become a dominant method: 1) It lacks the pedigree or purity of material developments such as concrete, with which it is both dependent upon and developmentally intertwined; 2) Similar approaches to construction systems and strategies such as "Chicago frame" or the residential adaptation of "balloon frame" construction technologies are far more integrated with architectural history and thus influence, and finally, 3) Tilt Wall construction has more in common with parochial methods of cultural procedures such as barn raising, at least considered as a commercial manifestation of technique rather than any kind of communal activity. Tilt Wall, at least, has in common with barn raising its conception as a unitized system of efficient building. In some ways, the fact that Tilt Wall has persisted to become so prevalent as a building method without the support of nearly any scholarship speaks to a deeply-embedded media / historical narrative that repeatedly privileges cultural activities like barn raising over commercially applied vernacular techniques like Tilt Wall. Thus, a surprisingly small number of sources record scant information regarding detailed development of Tilt Wall construction as a method.[i] Rather, it has evolved to its current state organically along several parallel vectors into a uniquely American approach to construction.

Origins

Tilt Wall's known history begins in the early 1900s with Colonel Robert H. Aiken, at one point an army engineer, pioneering the technique of pouring concrete slabs and tilting them into place as rifle range target abutments. Aiken, given the patents he filed in relation to ferroconcrete reinforcing techniques and tilting devices appears to have been possessed of a degree of assiduity in his adopted field of construction. Several articles spanning from 1908 to 1917 in trade publications of the day such as *Keith's Magazine on Home Building*, *Concrete Era*, *Concrete Engineering* and *Cassier*, cover Aiken's efforts in the building of structures using Tilt Wall as well. Much of what is known of Aiken is found in a paper titled, "Monolithic Concrete Wall Buildings—Methods, Construction and Cost" which he produced for presentation at the Fifth Convention of the American Concrete Institute (ACI) held in Cleveland, Ohio, on 13 January 1909. The summary of proceedings show he was not in attendance and a W.S. Abbott presented the work in his absence. Given that Aiken pioneered the method and that this short paper constitutes the known origin of its history, a brief overview is in order.

The first few paragraphs set up the introduction to what Aiken describes as a "new mechanical method of erecting buildings with concrete", supported by facts and statistics indicating the increased availability of concrete in the states from 1885 to 1906. To provide context to his declarations, it is worth noting Kenneth Frampton has recorded, in relation to a description of the American pioneer of reinforced ferroconcrete Ernest L. Ransome (author of *Reinforced Concrete Building*, 1912), that after 1895 the use of concrete in the United States becomes less dependent upon the importation of European cement, and thus establishes hegemonic status as a construction material nearly overnight. Aiken, prone to perspicacity, secures the brief introduction with the declaration of "a New Age of Concrete". This is evidence of the intoxicating times during which he practiced: that seam between the last vestiges of the Industrial Revolution and the rapid technology infusion of the pre-modern one taking its place. He gives a brief description of what will have to suffice as the invention of Tilt Wall construction that involved the pouring of five-foot-tall panels lying on the ground then raised to the vertical via a "derrick" (he anecdotally annotates as "horse drawn") onto an existing foundation then secured with pre-placed reinforcing bars. The function of this "tilted up" retaining

Colonel R. H. Aiken

Camp Perry construction

Camp Perry construction

wall was to be the above mentioned rifle range target abutment. He gives no exact date for this installation but as the paper was presented at the 1909 conference and he had built a farm factory building just prior to a 1906 church, both with the tilt up method, it must be assumed to be circa 1905, just prior to the establishment of what would become Camp Perry in 1906. According to his paper's stated chronology, he then translated the procedure to that of producing freestanding building walls and eventually buildings (from 1907) which comports with the publication dates in the magazines outlined above. That is not to say he was not working much earlier on the development of the components of a reinforced concrete system. Patents filed as early as 1903 and granted in 1908 show him working on topics such as the "process of making iron from the ore" and "making iron from the sulfides", both presumably dealing with reinforcing bars for concrete, but certainly obtaining to his investigation of a system of building in a larger sense.

Those early buildings include his self-proclaimed first fully Tilt Wall building, the 1906 concrete factory for producing targets with target frames, on Aiken's own farm.[ii] This project was immediately followed by a Methodist Church, located in Zion, Illinois,[iii] approximately one hour north of downtown Chicago. In a history of the church from 1837 to 1958, reference of the method is provided:

> Construction of the new edifice was done by Col. Robert H. Aiken, inventor of a system of building with concrete. Sections of the walls were poured in forms on the ground and, when hardened, were set into place, scrolls and flowers, beautifully executed in concrete, adorned the front of the building and five cherubim smiled above the door.

Aiken next relates having constructed a building he described as having attracted great attention, the Mess Hall at Camp Perry, Ohio.[iv] Here he had developed or at least employed a steel tipping table to replace the derrick-lifting method he had started with. It is unclear as to whether the Tilt Wall concrete structure or the method that it was erected with caused the interest. Arguably, the tone of his paper and his occupation as an engineer pertain to Aiken's innovations as more focused on the concrete reinforcing, the technology for actually tilting the panels into vertical position and the means and methods of efficiency in the process, than in the aesthetic design potential of the technique itself. In a patent posthumously filed by

Jannette K. Aiken in 1928, exact drawings of the lifting platforms described and partially illustrated in his 1909 paper in relation to the Camp Perry Mess Hall building are provided. In the paper, he conveys a kind of tipping table operated by hydraulics and actuated by a five-horsepower motor. It appears that the photographs provided in the paper, of the Mess Hall during construction, are of the same design he eventually desired to patent; however, no motor is illustrated or described in the patent filing. Interestingly, Aiken himself filed for a patent in 1908 (No. 889,083), concurrent with or just prior to the Mess Hall project, related to reinforced concrete construction as a general topic, or more particularly several "improvements" to the method itself, but not of the lifting tables themselves.

While the paper promises to deal with monolithic concrete wall buildings by title, Aiken, it is clear, saw ferroconcrete as a process and sought to innovate in the coming together of the material and the method of its utilization. In this way, he saw and indeed attempted to claim innovative novelty in them as a synthesized system. This "systems" approach to concrete construction is further underscored by a related patent issued to Aiken in 1905 also focused on reinforced concrete construction, in particular, methods for site or precast floor-beam pockets to be designed into wall panels. Seen in light of the 1903 work in iron derivation from sulfides and ores, innovations of the tipping table and general specification of reinforced construction methodologies as a whole, his efforts to commercialize and commoditize a system put him in the intellectual company of ferroconcrete innovating predecessors Joseph Monier and François Hennebique, both of whom developed and commercialized proprietary techniques in material properties and systems. He then ultimately makes claim in his paper that:

> More than half a mile and fifteen different structures of one kind or another, in five states, and, up to this time, of one and two stories only, have been reared of this construction.

The balance of his proposition hinges towards visionary projects for housing developments to be built using the Tilt Wall method and descriptions of how the technology, if it may be called that during the time of his conception of it, (more on this in Chapter 5) could produce buildings of many stories. Aiken's paper, presented as the last of a Thursday evening session, summarizes what is recorded of his formal discourse on the Tilt Wall system. The balance of his argument was to be made through his mildly prosaic attempt at commercial activities.

FIG. 10.—REAR VIEW OF WALL IN VERTICAL POSITION.

Camp Perry construction

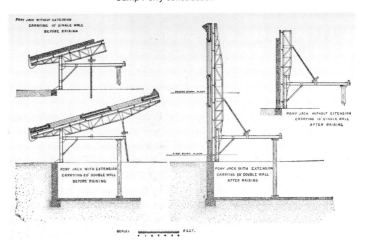

FIG. 5.—WALL RAISING JACK. LIMIT OF HEIGHT, PONY SIZE 20 FEET, STANDARD 30 FEET, JUMBO 40 FEET.

Aiken's Steel Tipping Table

There is no evidence he ever built or executed anything more than what he describes in the paper, but ground work had been established for several developments. It is known that Aiken at some point formed the Aiken Reinforced Concrete Company. This is established in the negative by historical records of court documents related to legal proceedings against the company filed in June 1919. However, the documents that refer to the company's performance on a brick building structure, combined with architect Irving Gill's later purchase of the rights to the technology or the actual Aiken company itself, are unclear. This makes the formation of his company and its termination slightly ambiguous and perhaps not as important to his legacy as the technique itself. Thus it is uncertain whether Aiken achieved much in the way of additional commercial success with the technique. Having only the above accomplishments capable of catapulting him into the canonical history, Aiken in most ways remains a secondary adumbrated figure, and so his understanding of the implications of his invention of Tilt Wall, concomitantly remains so as well. He is precluded from any major architectural history from Sir Bannister Fletcher and Sigfried Giedion through Reyner Banham and including Frampton. Indeed he does not warrant mention in Edward Ford's *The Details of Modern Construction*. Like many minor figures that *do* receive mention in these texts, the context in which he was speculating about new construction technology is as important as who he was. That context is defined primarily by the fact that it was concrete he was working with in his drive to efficient methodology; the critical historical period in which he was speculating about the "new" in construction, and that he was in the orbit of Chicago and its pioneering of construction innovation, thus by inference his interest in efficient and new construction methods.

Concrete

While the history of the development of concrete as a major construction material reaches as far back as pre-Roman developments, it is ferroconcrete or iron / steel reinforced concrete that were Aiken's area of interest, invention and research. But why was it concrete that Aiken, an army engineer and commercial aspirant, had become interested in? Why in particular a method of erection of concrete structures rather than focused development of his technical patents in ore and metal sulfide processes? The technical history of concrete and ferroconcrete is well developed and documented. Giedion concluded as much in *Space, Time and Architecture* as he chose to focus on the convergence of ferroconcrete and architecture, a point at which it is easy

26

to place Aiken and his Tilt Wall innovation. Paraphrasing from overviews by Giedion and Banham, a concise timeline of just how concrete came to the forefront of construction technology and architectural investigation is derived:

> In 1774, some five years prior to the erection of the first cast iron bridge and for the first time since the Romans, English engineer John Smeaton used a mixture of quicklime, sand clay and crushed iron slag; otherwise concrete, to bind the stones in his reconstruction of the Eddystone lighthouse. Joseph Aspdin developed Portland cement as a hydraulic, or cement that hardens under water, in 1824. This is followed by innovations in using concrete as filling between iron girders for floors and finally tie-bars woven into the concrete pioneered still by English engineers, in this case William Fairbairn in 1844. The science of the interaction between the concrete and the iron reinforcing took form nearly fifty years later in French circles, passing iteratively through developments by several innovators. François Coignet, T.H. Monnier, G.A. Wayss and finally German engineers Nuemann and Koenen refined the predictability of differential stresses between the iron and concrete materials to form true ferroconcrete. Completing this international consortium of development, it was American Ernest Ransome who used the material the first time.

The work of François Hennebique is perhaps of dominant importance in emphasizing developments and research that may have been available to and influential upon Aiken. Hennebique and his work are well documented; what is of importance here is his unique approach to the commercialization of his research and methods. Hennebique, a builder by training, invented a system of jointure using reinforced concrete, then innovated by conceiving of the solution as a system of construction that came under patentable form. He was just as interested in an approach to using the new material as he was in its theoretical development, as were many of his peers, but he solved it and commercialized it. Many of his contemporaries endeavored to do so as well. Monier developed a system, as did the engineer Paul Cottancin, but no one had previously experienced the success Hennebique achieved. What he ultimately invented was a reinforced concrete frame system, a simple skeletal matrix that had unlimited potential and also untapped architectural implications. In the hands of others, such as Auguste Perret, Max Berg. Robert Maillart, Eugène Freyssinet and ultimately albeit in highly evolved form Le Corbusier's in the Maison Dom-ino, it became one that was critical to architectural aesthetic development. It was also a system that as a structural frame, left ferroconcrete unaddressed as an external wall system, thus Aiken's opportunity was established, whether he was conscious of it or not. It is clear, however, that by 1912 reinforced concrete frames were in common, if not ubiquitous use, even as Aiken chose to pursue utilizing reinforced concrete slabs poured horizontally and tilted into place as load-bearing structural walls. Concrete was for Aiken the cutting-edge technology of the day. As to why he focused on the methodology of constructing with concrete rather than its aesthetic potential or its plastic possibilities, the answer is more nuanced; it appears to be a fusion of his engineering capability distorted by market opportunity and fueled by a keen sense of originality that drove him.

Times

As mentioned above Aiken's 1909 paper was preceded by several patents filed as early as 1903 and his construction of several tilt up buildings. By way of context in innovation, in 1909, Einstein had formulated his Special Theory of Relativity, Bakelite (the commercial precursor to plastic) had entered commercial production, the Ford Model T car was in mass production, Picasso and Braque had already exhibited their Cubist works, and, closer to architecture, Frank Lloyd Wright had completed the seminal Robie House just as Charles McKim, purveyor of the classical in American architecture, died. It was a time of massive technological transformation and historical evolution into the modern age. Like any transitional time, the old often maintained an existence alongside the new, and it is at the cleft between the old and new that Tilt Wall construction was born. A refined approach to mechanically systematizing construction as a commercialized business venture benefitted from the hubris of the dying Industrial Revolution giving way to the establishment of Modernism. The last of the former fueled the halcyon days of the latter in numerous ways. Perhaps the most influential on Aiken may have been the formalization of industrial production promoted and popularized in the establishment of The Great Exhibitions of the Works of Industry of All Nations. Architectural historian Leonardo Benevolo structures a great deal of his first volume of the *History of Modern Architecture* on the age of the great universal exhibitions. The purpose of these uniquely global expositions also bears directly on the event that may have brought certain direct technologies, if not the notion of commercial innovation, to Aiken's doorstep some years later. According to Benevolo political developments supporting the possibility of international trade inspired the need to market technology and products at

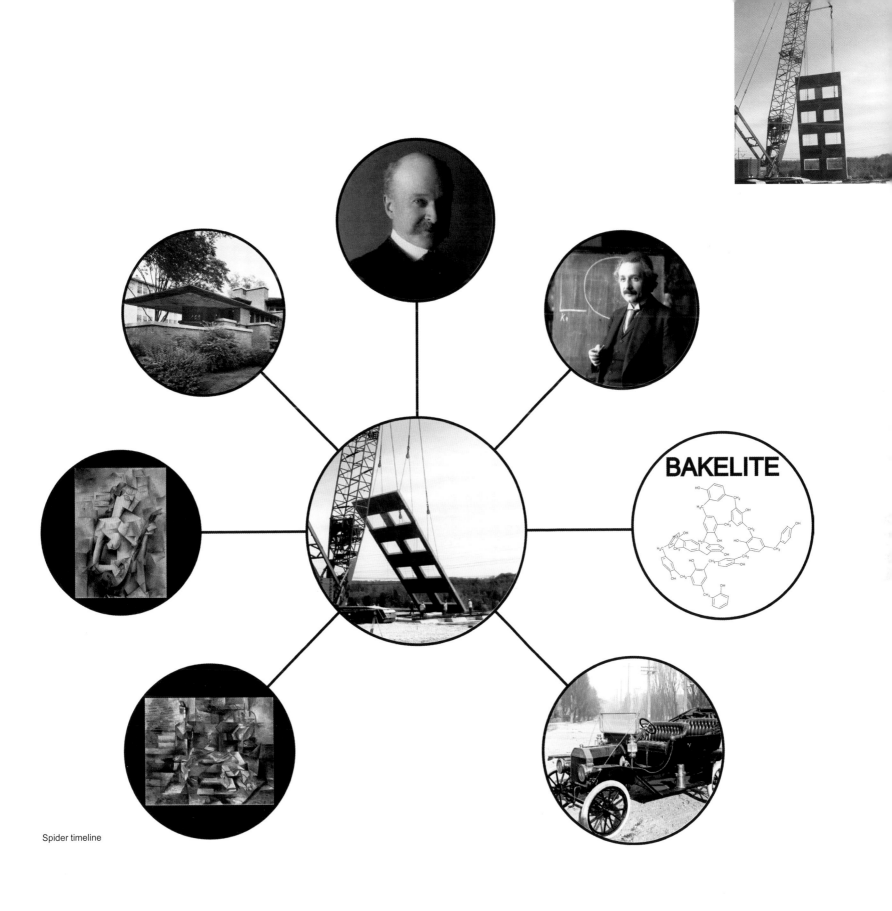

Spider timeline

The Crystal Palace, London

The Crystal Palace, building interior

Champs de Mars, Galerie des Machines, exterior

Champs de Mars, Galerie des Machines, interior

a maximum scale. Forums for the display of manufacturing processes, goods, inventions and other national production were required to solicit international trade, and this was precisely the purpose of the Great Exhibition, held in 1851 just outside London in The Crystal Palace. The building not only became famous and influential, but the tradition it began ever increased in scope and range. With everything from machines to materials, goods and wares on display from nearly every nation with a developed economy, the roots of the vision that became modernity were established. The 1867 Parisian version of The Crystal Palace, on the Champs de Mars, was a building with several galleries, including one named the Galerie des Machines. It was here that Monier exhibited his now-famous reinforced concrete pots, the progenitors of ferroconcrete as an architectural building technique. Subsequently, and catalyzing the notion of a systems approach to construction technology presumably brought on by the ever-increasing division of labor between architects and engineers begun in the early 19th century, Monier, Hennebique and others designed and laid their claim to proprietary systems for concrete construction techniques and technology, presumably inspired by what they experienced in these international gatherings. As relates to abstract influences on actors such as Aiken, the great halls themselves in their innovation, particularly The Crystal Palace, reified the height of the Industrial Revolution in what has been described as "not so much a form as it was a building process made manifest as a total system". This serves as a precursor to the mindset Aiken may have had, in creating the erection of concrete, its design and its architectural potential as a system in and of itself.

It is through the universal exhibitions, culminating in the 1893 American contribution to the movement of international gatherings, the World's Fair: Columbian Exhibition, that any new innovative construction or related technology would have been available to influence a polymath like Aiken. Mt Zion, the locus of Aiken's early production, and Winthrop Harbor, Illinois, his residence, are approximately one hour north of Chicago, and while there is no evidence he attended the year-long Columbian Exhibition, it is not a stretch to conclude he was aware of it at the very least. An examination of the planned contents of the exhibition program reveals no clear "construction" techniques on display; however, there was a plethora of global production techniques on view in numerous disciplines. The exhibition was not a success seen from the position of advancing architectural Modernism, nor advancement in the architectural innovation of Chicago's practitioners; indeed in some ways it truncated with

Classicism the advent of Modernism for a short period. Yet its opportunity to put America on the global stage contributed to the transformative environment of Chicago's architectural development.

Chicago

It is not a stretch to speculate that the force of these developments could have borne either directly or abstractly on Aiken given his practice in the orbit of Chicago in the early 20th century. The Columbian Exhibition of 1893 certainly brought the spirit of the tradition of international cross-pollination right to Aiken's back yard. Chicago was, more so than any other American city, prone to being on the forefront of the development, employment of and experimentation with commercial manifestations of new technologies related to construction. From the skyscraper, to the elevator, to the Chicago caisson, the "Chicago school" would have been an undeniable influence on any sentient practitioner of the building trades in these technologically fertile times. Indeed Sullivan, according to Frampton, observed that in order to practice architecture in Chicago one had no choice but to master advanced methods of construction. Of course, Sullivan stood outside of many of the protagonists of mere practice as a theorist and advocate of a unique American architecture. Aiken intersects, at least by common training and as contemporary, the grandfather of modern Chicago architecture William Le Baron Jenney. Jenney was trained in the Army Corps of Engineers during the Civil War much the same as Aiken was during the Spanish-American War. Jenney died in 1907, thus overlapping Aiken's active practice life. More importantly to the legacy of pre-Columbia Exposition Chicago and evidence in support of the influences that may have pressed Aiken to innovate in Tilt Wall technology, Jenney was the first in a long line of important Chicago architects of the same era. Major innovators in both practice and technology such as Daniel Burnham, John Root, William Holabird, Martin Roche, and, for a lesser overlap, Dankmar Adler, were active concurrently with Aiken. And while the exhibition may have brought a spirit and encouragement of entrepreneurialism and specific technological innovation within Aiken's orbit, the invention of the steel-frame skyscraper and, concomitantly, the elevator alongside the structural advancements of the Chicago Auditorium Theatre and like buildings were bound to inspire downstream local practitioners much the same way they caught the attention of leading designers in other American cities of the time. Certainly the unique commercialized cocktail of commerce and design that characterized the city would have been an opportunity for the Aikens of the day.

Irving Gill

La Jolla Women's Club and Clubhouse under construction

La Jolla Women's Club and Clubhouse

Integration into Modernism

Beyond Aiken, early significant practitioners of the Tilt Wall technique were limited, but as a matter of rearguard action it must be noted that while Thomas Edison created a cluster of reinforced concrete houses in 1908 at Union Village, New Jersey that still exist today,[v] they were form cast and not, as is often claimed, Tilt Wall construction. Interestingly, Thomas Fellows advocated a similar interest in tilting walls into place and in or around 1910 utilized a variation on Aiken's method for a low-cost demonstration house. His method, described in an article in *Southern Contractor and Manufacturer* magazine, used an eyebolt system in which rods tied the panels together at the vertical joints, later to be grouted. Interestingly the article describes a crane that lifted the panels into place, a precursor of the perfection of the method many years later. As to whether Fellows had knowledge of Aiken's system, we can only speculate. It may, for example, be a case of what physicist Freeman Dyson has termed "complementarity". Roughly defined, it is the simultaneous unconnected invention of similar theories. But that Aiken's system was known on the West Coast, we know to be a fact.

Tilt Wall construction briefly interested California regionalist architect Irving Gill, who experimented with it in several projects and as a business. Gill was inveterately interested in innovation and materiality in his work. He combined this with a passion for the craft and making of his buildings, often creating integrally cast doorframes for his early cast-in-place houses, along with developing shop-fabricated wall panel systems and approaches to heating / cooling system approaches. Frank Lloyd Wright's son, Lloyd Wright, worked for Gill, as did Gill's nephew, Louis Gill, at a time in San Diego in which contractors built all but the public buildings and larger residences.[vi] Irving Gill was unique in his interest in overcoming the exclusion of architects from everyday building types; the symmetry with current or contemporary trends is stunning. With the purpose of building more economically and more quickly in order to compete with the then dominant contractors, Gill purchased the rights in 1912 to the by-then bankrupt Aiken Reinforced Concrete Company.[vii] Gill first used it on the Banning House in 1913 to the amazement of neighbors, according to Sunset magazine at the time.[viii] In 1914, Gill partnered with Louis to create the Concrete Building and Investment Company in order to build low-cost housing. There is, of course, a parallel in that Aiken also utilized

the method for low-cost (in his case barracks) housing and that both men saw in Tilt Wall construction the intersection of architecture and business. In some ways, they were early John Portman-types, with Gill exhibiting a modern developer sensibility. Gill has been accepted into architectural history more for his early adaptation of Modernism in regionally inflected work, significant in his opus the La Jolla Woman's Club for which he utilized the Tilt Wall method to conceive of and construct just prior to forming the building company. The company ultimately failed at great cost to Gill. Suffice to say, it has been the Tilt Wall industry that has revived research into Gill on this subject, not Gill who pushed it forward in history.

Irving Gill also gave Tilt Wall one additional dimension to its pedigree in high Modernism by introducing Rudolph Schindler to the method. While Gill had the aesthetic implications of Modernism in mind with construction playing a secondary role in his work, Schindler's work is defined and characterized by his search for the moral depths of Modernism; the relationship to form and craft were important areas of his investigation. Lloyd Wright, with whom Schindler had worked during his time at Taliesin in Frank Lloyd Wright's employ, introduced Schindler to Gill and his work. In fact Schindler recorded some of Gill's Mission Revival style work while on a tour in California in the early 1920s and also visited Gill in his office. It is there that Schindler presumably became aware of the Tilt Wall method. He employed it in the design of his home and studio at Kings Road, directly across the street from one of Gill's most significant residential projects, the Walter L. Dodge House in Los Angeles. Schindler's use of the method in the Kings Road house was, at best, secondary, but still innovative. He poured one-story walls—that would contain the main rooms of the connector between wings of the house—on the slab and utilized a block and pulley system to tilt them into place vertically. The edge of each wall, at the joint between panels, was left as a gap into which a thin glass "connector" articulated each panel joint.

Rudolph Schindler

Kings Road House under construction

Lacunae—contemporary diaspora

After Tilt Wall's flirtation with high Modernism, it sank into anonymity for several decades as the war retarded the commercial construction industry. However World War II and its economic aftermath sponsored a need to build quickly and economically. Surviving the war as a technique but shedding its high modern pedigree, Tilt Wall was revived as the way to

Kings Road House

build large-footprint, big-box projects. Experimentation on its use in low-cost housing, conducted almost in ignorance of its history, was flirted with in the late 60s and 70s, and documented in trade publications from the mid 1960s to the early 1970s. The invention of the mobile crane in the post-war construction market facilitated the lifting of ever-larger panels on nearly any site. Its inherent economies driven by its speed of erection coupled with its efficiency in creating large boxes—the slabs of distribution centers and warehouses being proportionally strategic as casting beds for the wall panels—resulted in the method being relegated, for the most part, to "big dumb box" architecture as the 20th century came to a close. With its increased employment as a low-cost technology came the ubiquity that has come to characterize both its market depth, in terms of potential buildings types, and its untapped architectural design potential that define its opportunity to architects today. Contemporary "starchitects" such as Steven Holl along with regional innovators Mack Scogin Merrill Elam and Rand Elliott have shown both interest and innovation can be achieved using Tilt Wall construction, but they also illustrate that much of the work in tapping its true potential is still in front of the profession.

i   Tilt Wall construction receives only notational mention in Edward Ford's *The Details of Modern Architecture* (MIT Press, Massachusetts, 2003) and is not addressed in Frampton's recent *Studies in Tectonic Culture* (MIT Press, Massachusetts, 1995), arguably two of the more dominant history / theory works that circumscribe constructions role in architecture.

ii   This was self-proclaimed by Aiken according to Maura Johnson in "Tilt-up Pioneer: Robert Aiken Developed Tilt-Up Construction Nearly 100 Years Ago", 1 August 2002.

iii http://www.concretecontractor.com/tilt-up-concrete/construction-history/.

iv The Mess Hall building lasted nearly 100 years until it was demolished in 2001 after suffering damage in a 1998 tornado.

v   http://www.concreteconstruction.net/images/Tilt-Up%20Pioneer_tcm45-590019.pdf.

vi "Esther McCoy from Five California architects included in Marvin Rand's book *Irving J. Gill: Architect 1870-1936* (Gibbs Smith, Utah, 2006).

vii Ester McCoy states it as equipment purchased from the army.

viii Ibid.

" ... it sank into anonymity for several decades ... "

A tilt-up construction project begins with job site preparation and pouring the slab. During this phase of the project, workers install footings around the slab in preparation for the panels.

They provide the panels' exact shape and size, doorways and window openings, and ensure the panels meet the design specifications and fit together properly. Next, workers tie

The slab beneath the forms is then cleaned of any debris or standing water, and workers pour concrete into the forms to create the panels. This is where tilt-up construction gets

The size of the crane depends on the height and weight of the cement panels, but it is typically two to three times the size of the largest panel. The crew also attaches braces to

sets it into place. They connect the braces from the tilt-up panel to the slab, attach the panel's embeds to the footing, and disconnect the cables from the crane. The crew then

The crew then assembles the panel forms on the slab. Normally, the form is created with wooden pieces that are joined together. The forms act like a mold for the cement panels.

in the steel grid of reinforcing bars into the form. They install inserts and embeds for lifting the panels and attaching them to the footing, the roof system, and to each other.

EXCURSUS A
How Tilt Wall Works

The Technique Defined

its name. Once the concrete panels have solidified and the forms have been removed, the crew connects the first panel to a large crane with cables that hook into the inserts.

the tilt-up panel. The crane lifts, or "tilts up," the panel from the slab into a vertical position above the footings. Workers help to guide the concrete panel into position and the crane

moves to the next panel and repeats this process.[i]

i   www.tiltup.com-commercial construction articles

CASTING SURFACE

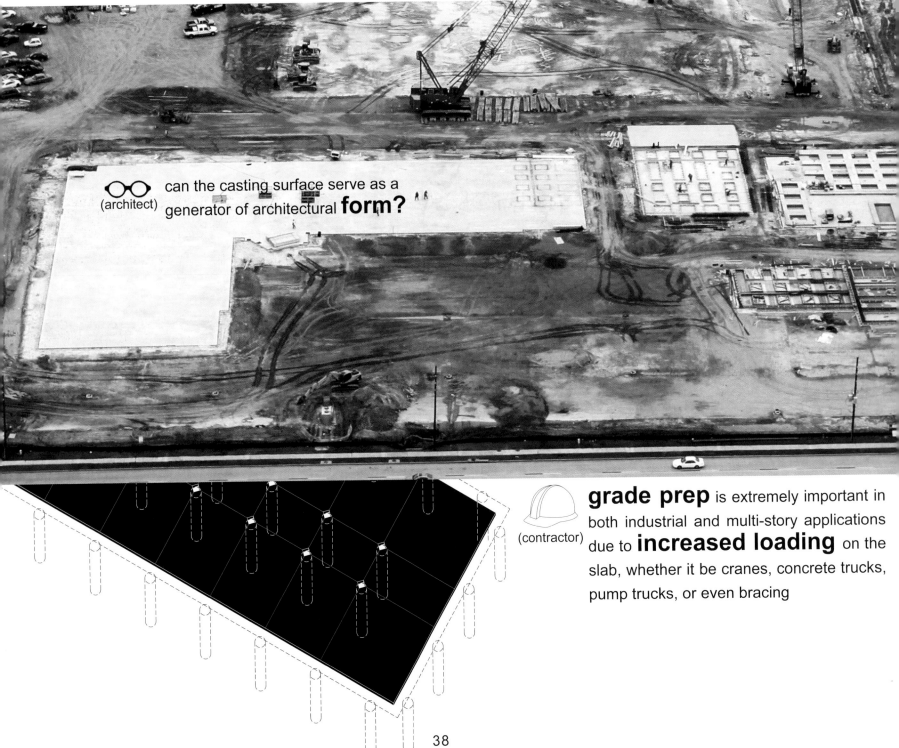

(architect) can the casting surface serve as a generator of architectural **form?**

grade prep is extremely important in both industrial and multi-story applications **(contractor)** due to **increased loading** on the slab, whether it be cranes, concrete trucks, pump trucks, or even bracing

flat casting surface (minimal ponding water after rain) with hard-trowel finish

(engineer)

PONDING

VS

future concrete slab

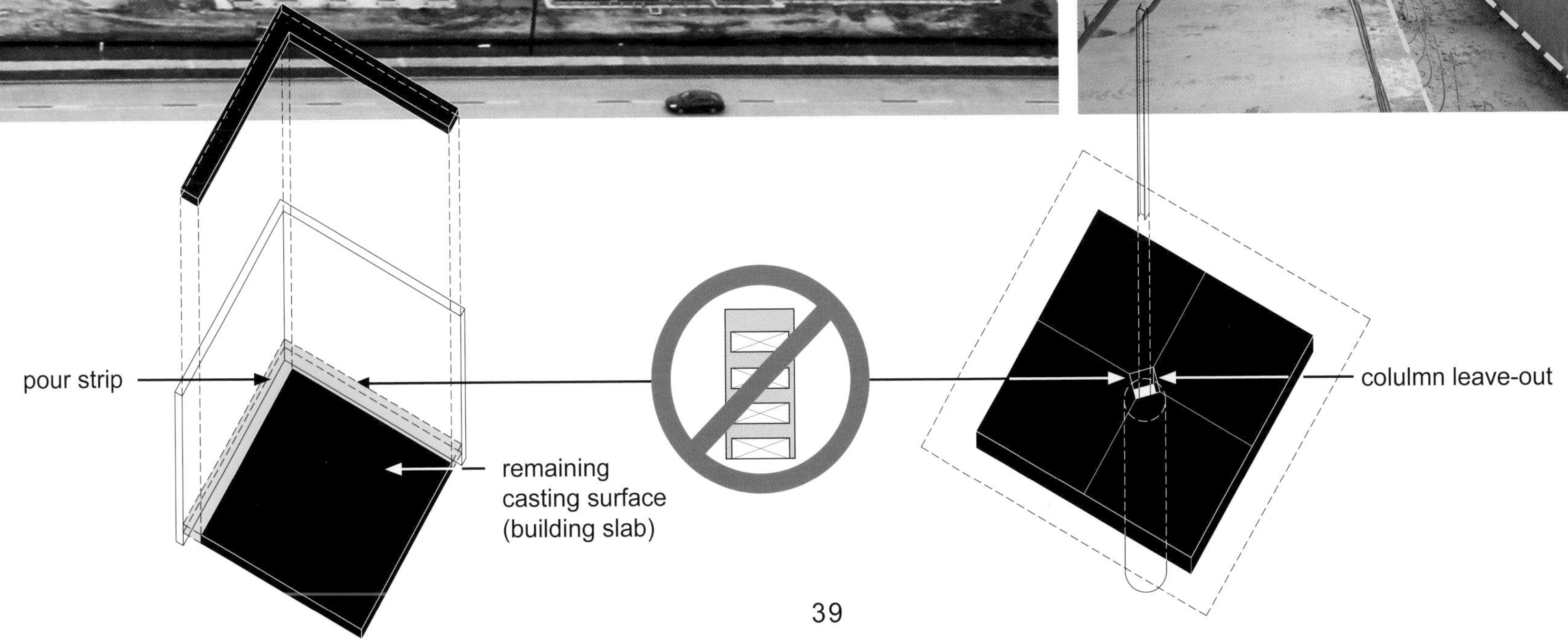

pour strip

remaining casting surface (building slab)

colulmn leave-out

39

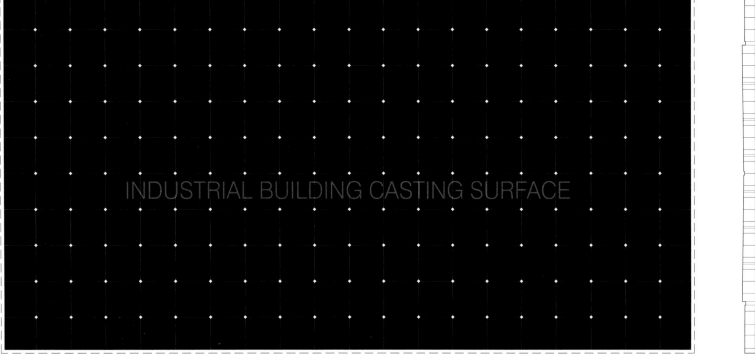

INDUSTRIAL BUILDING CASTING SURFACE

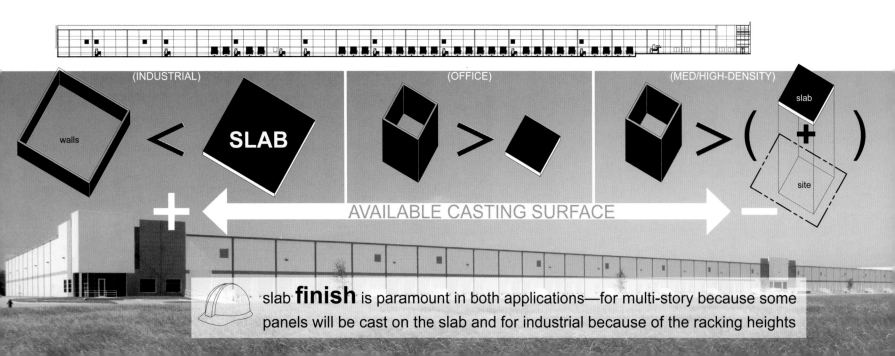

(INDUSTRIAL)

walls < **SLAB**

(OFFICE)

>

(MED/HIGH-DENSITY)

slab

> (+)

site

+ ←――――― AVAILABLE CASTING SURFACE ―――――→ **−**

slab **finish** is paramount in both applications—for multi-story because some panels will be cast on the slab and for industrial because of the racking heights

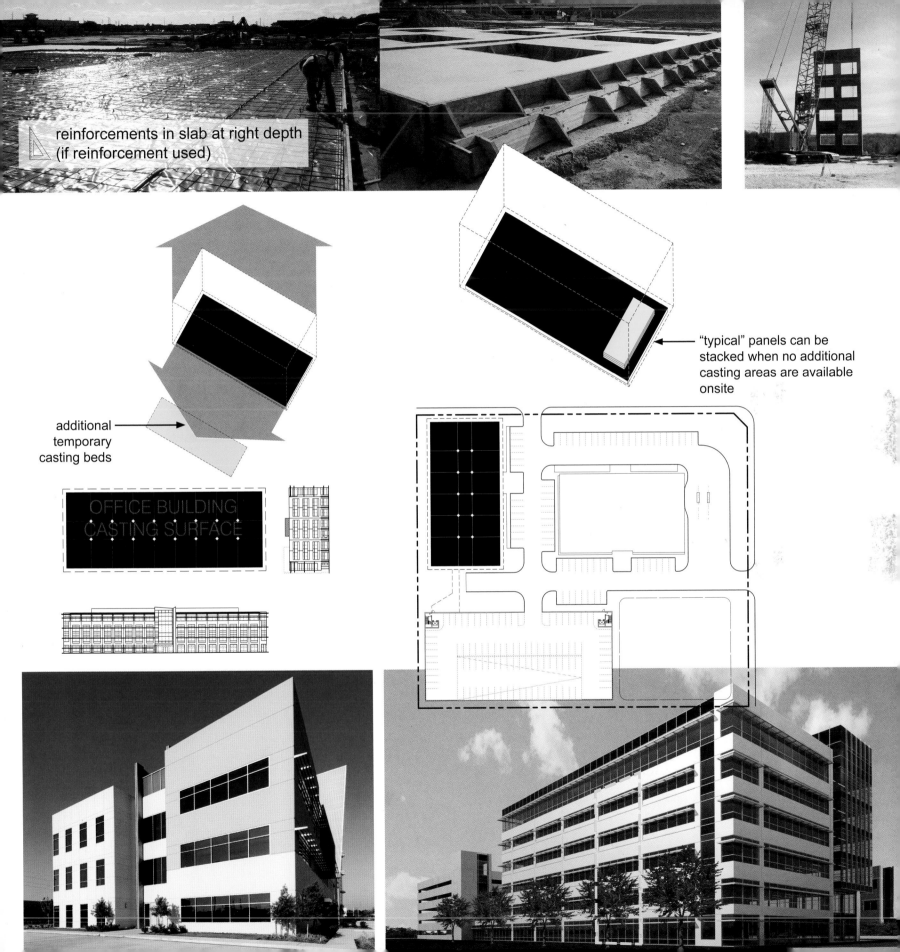

reinforcements in slab at right depth
(if reinforcement used)

"typical" panels can be
stacked when no additional
casting areas are available
onsite

additional
temporary
casting beds

OFFICE BUILDING
CASTING SURFACE

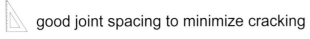 good joint spacing to minimize cracking

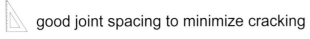 plate- or bar-dowels at cold joints

⚆⚆ alternative casting surfaces onsite can create varying textures on panels

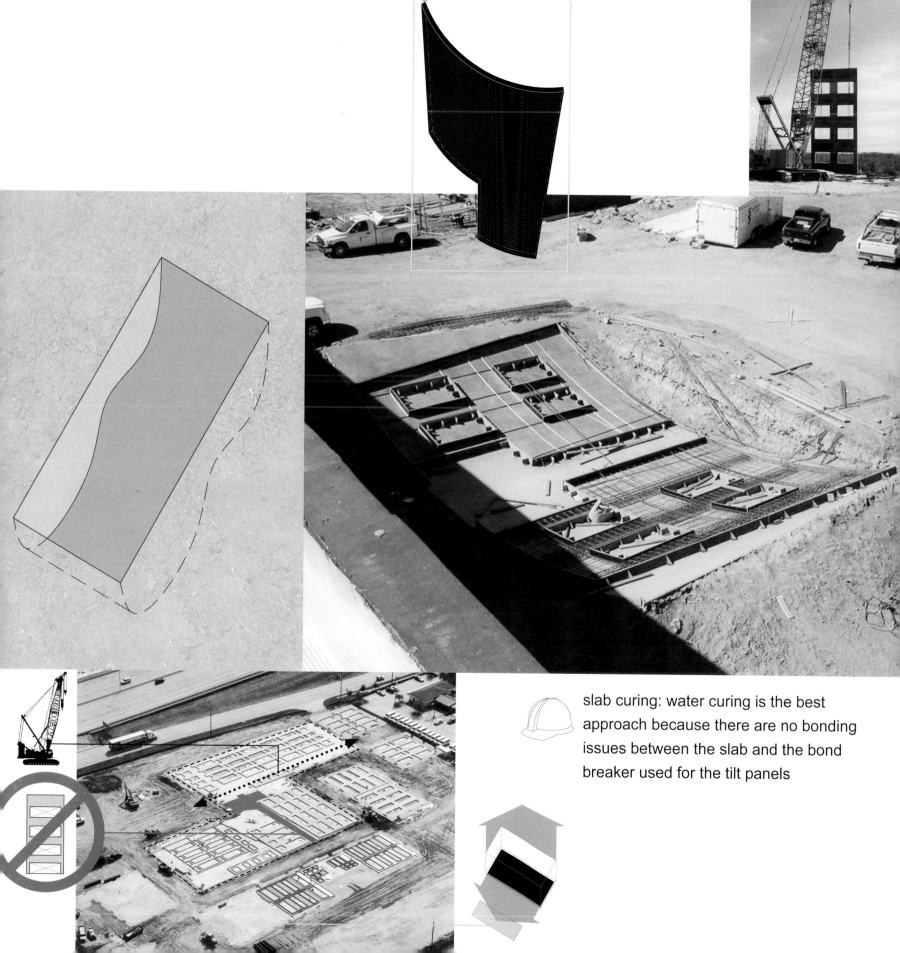

slab curing: water curing is the best approach because there are no bonding issues between the slab and the bond breaker used for the tilt panels

F O R M S

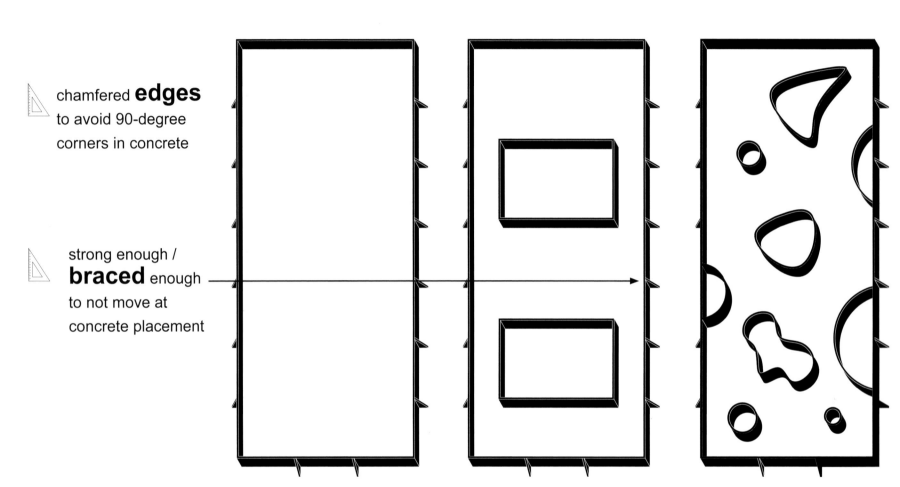

 chamfered **edges** to avoid 90-degree corners in concrete

 strong enough / **braced** enough to not move at concrete placement

form material: high-quality material is important to ensure **crisp lines** in openings and slab-edges

bracing of forms: ensure proper bracing of forms to prevent blow-outs when pouring panels—larger panels create signifcant force on the forms when placing concrete in the panels

(allowable reveal / formliner depth)

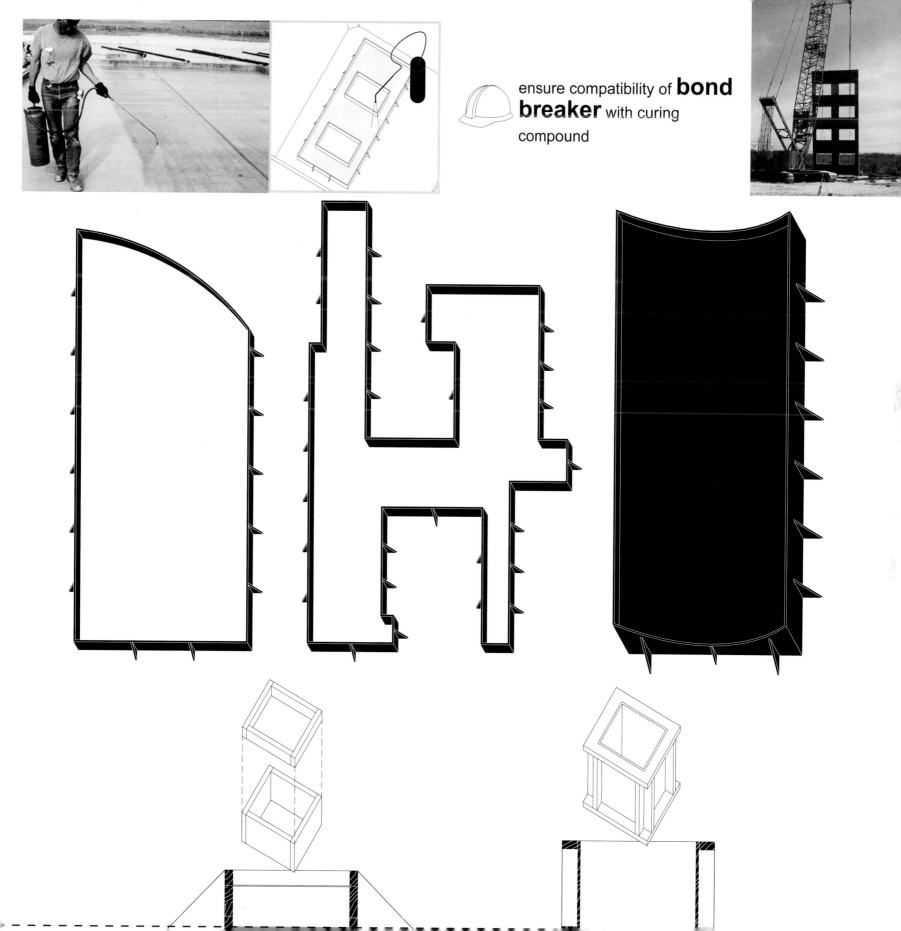

ensure compatibility of **bond breaker** with curing compound

straight, square, and plumb

 openings and overall panel size at correct dimensions

 caulked / sealed to avoid concrete-paste bleed-through

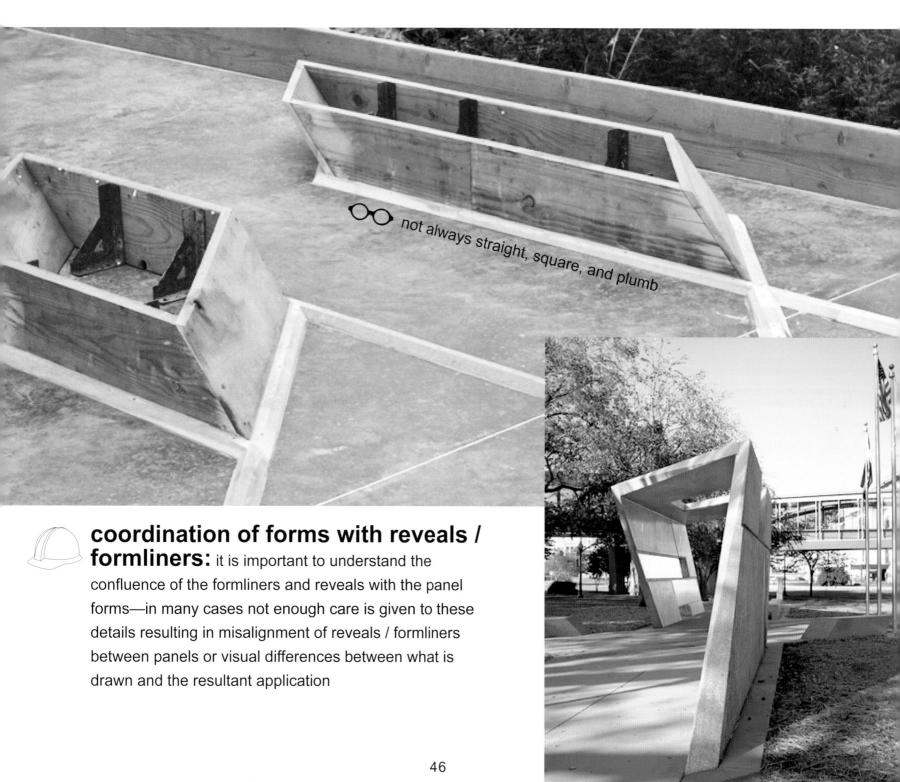

not always straight, square, and plumb

coordination of forms with reveals / formliners: it is important to understand the confluence of the formliners and reveals with the panel forms—in many cases not enough care is given to these details resulting in misalignment of reveals / formliners between panels or visual differences between what is drawn and the resultant application

 fully coordinated **panel drawings** to show the forming of panels and openings, reveals / formliners, reinforcing, lifting hardware, and layout on casting surfaces, etc.

REVEALS

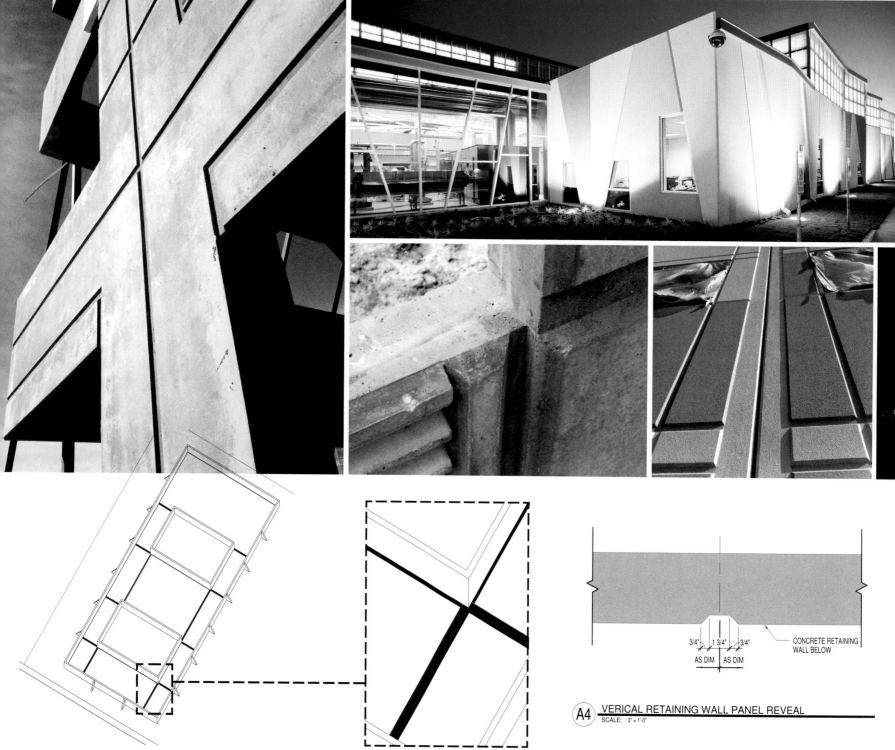

3/4"   1 3/4"   3/4"

AS DIM   AS DIM

CONCRETE RETAINING WALL BELOW

A4   VERICAL RETAINING WALL PANEL REVEAL
SCALE: 3" = 1'-0"

beveled edges

not deeper than design documents

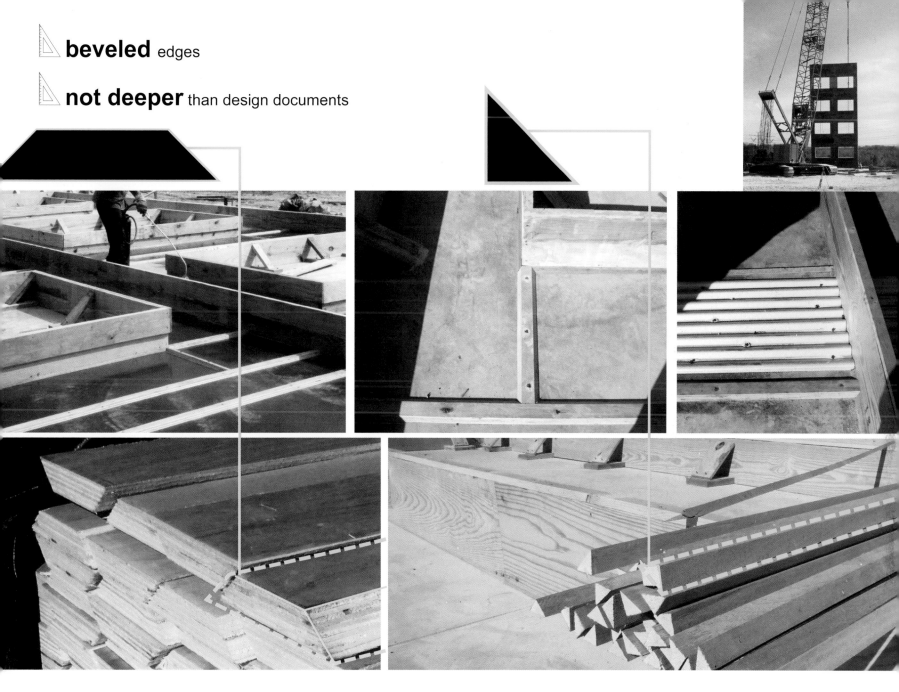

connection to casting surface: when nailing **reveals / formliners** to the casting surface, float all **hammer marks** and **seal** all nail **holes**

layout: always do a **full layout** of reveals on casting suface to ensure **proper alignment** and intersection

installation: **seal reveals** to casting surface to **prevent** the **bleeding** of concrete between the reveals and casting surface—this ensures **crisp** reveal **lines**

49

aligned from panel to panel

caulked / sealed to avoid concrete-paste bleed-through

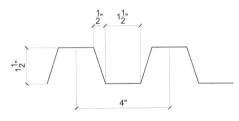
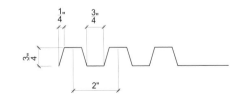
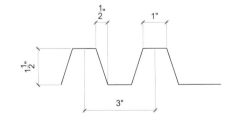
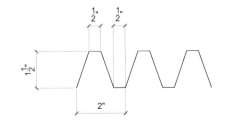
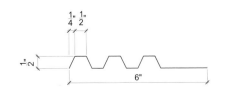
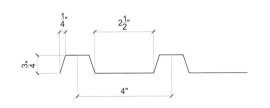

reveal material: should be **high-quality** material to **prevent weather damage** while stored onsite or in place

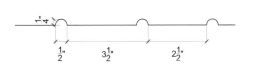

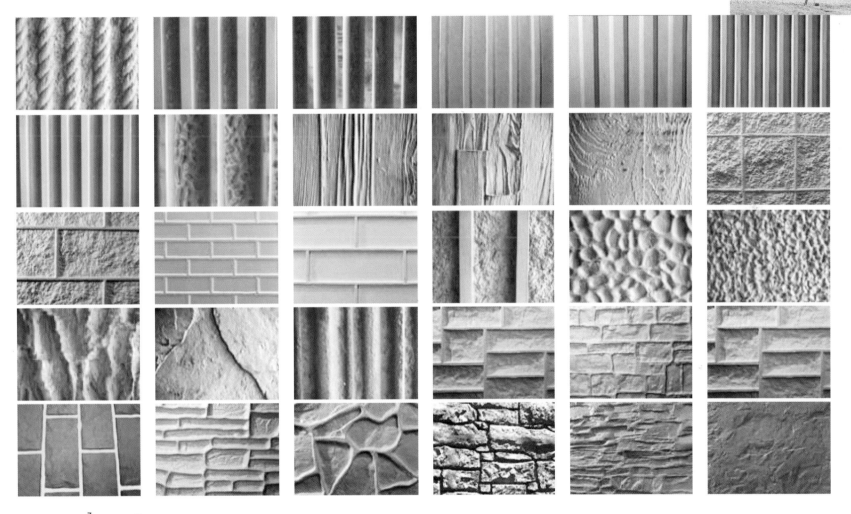

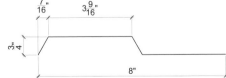

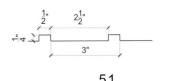

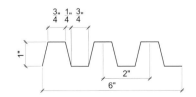

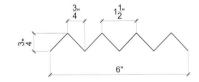

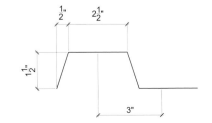

REINFORCING

reinforcing **too close to the surface,** or any other material at the surface, will result in imperfections in the panel finish or rust spots at the least

care should be taken to ensure that the reinforcing steel is in the **proper location** within the panel, as the greatest load on the panel occurs in the lifting process

minimal rust and free of bondbreaker / chemicals / oils

 diagonal bars at re-entrant corners

 enough **support chairs** so mat does not settle / deflect / displace when concrete is placed

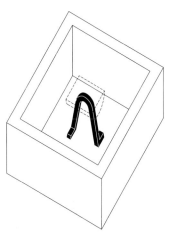

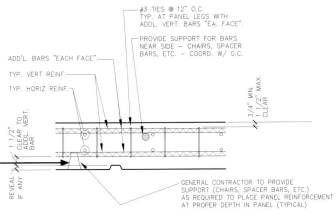

#3 TIES @ 12" O.C. TYP. AT PANEL LEGS WITH ADDL. VERT. BARS "EA. FACE".

PROVIDE SUPPORT FOR BARS NEAR SIDE – CHAIRS, SPACER BARS, ETC. – COORD. W/ G.C.

ADD'L. BARS "EACH FACE"

TYP. VERT. REINF.

TYP. HORIZ. REINF.

3/4" MIN. 1 1/2" MAX. CLEAR

1 1/2" CLEAR TO ADDL. VERT. BAR

REVEAL IF ANY

GENERAL CONTRACTOR TO PROVIDE SUPPORT (CHAIRS, SPACER BARS, ETC.) AS REQUIRED TO PLACE PANEL REINFORCEMENT AT PROPER DEPTH IN PANEL (TYPICAL)

PANEL REBAR PLACEMENT DETAIL – SECTION

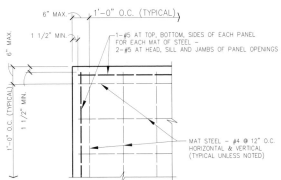

6" MAX. 1'-0" O.C. (TYPICAL)

1 1/2" MIN.

6" MAX.

1'-0" O.C. (TYPICAL)

1 1/2" MIN.

1-#5 AT TOP, BOTTOM, SIDES OF EACH PANEL FOR EACH MAT OF STEEL – 2-#5 AT HEAD, SILL AND JAMBS OF PANEL OPENINGS

MAT STEEL – #4 @ 12" O.C. HORIZONTAL & VERTICAL (TYPICAL UNLESS NOTED)

PANEL REBAR PLACEMENT DETAIL – ELEVATION

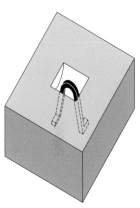

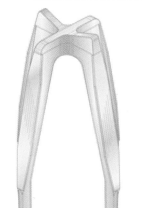   

 additional reinforcing should be, and often is, considered in the panels by the lifting engineer, not only at the lifting hardware, but around the openings and in the panel legs

 lifting inserts must be designed by the **lifting engineer**

INSERTS

 weather conditions often require that reinforcing steel is removed from panel forms and replaced—this should be a consideration in tying the steel

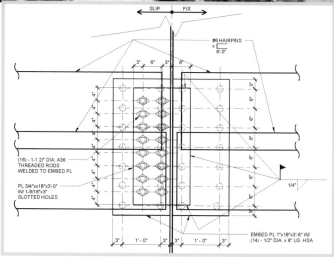

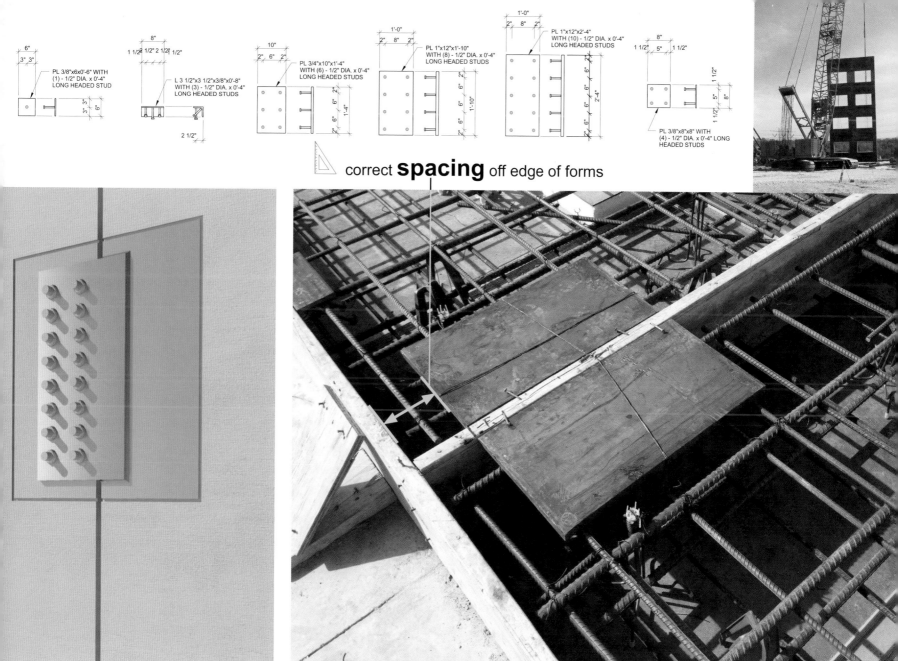

PL 3/8"x6x0'-6" WITH
(1) - 1/2" DIA. x 0'-4"
LONG HEADED STUD

L 3 1/2"x3 1/2"x3/8"x0'-8"
WITH (3) - 1/2" DIA. x 0'-4"
LONG HEADED STUDS

PL 3/4"x10"x1'-4"
WITH (6) - 1/2" DIA. x 0'-4"
LONG HEADED STUDS

PL 1"x12"x1'-10"
WITH (8) - 1/2" DIA. x 0'-4"
LONG HEADED STUDS

PL 1"x12"x2'-4"
WITH (10) - 1/2" DIA. x 0'-4"
LONG HEADED STUDS

PL 3/8"x8"x8" WITH
(4) - 1/2" DIA. x 0'-4" LONG
HEADED STUDS

correct **spacing** off edge of forms

coordinate the locations of lifting inserts with the overall design of the
building—locate in interstitial space on commercial product when possible

55

supplemental reinforcement as required by manufacturer or designer

 tied to main reinforcement

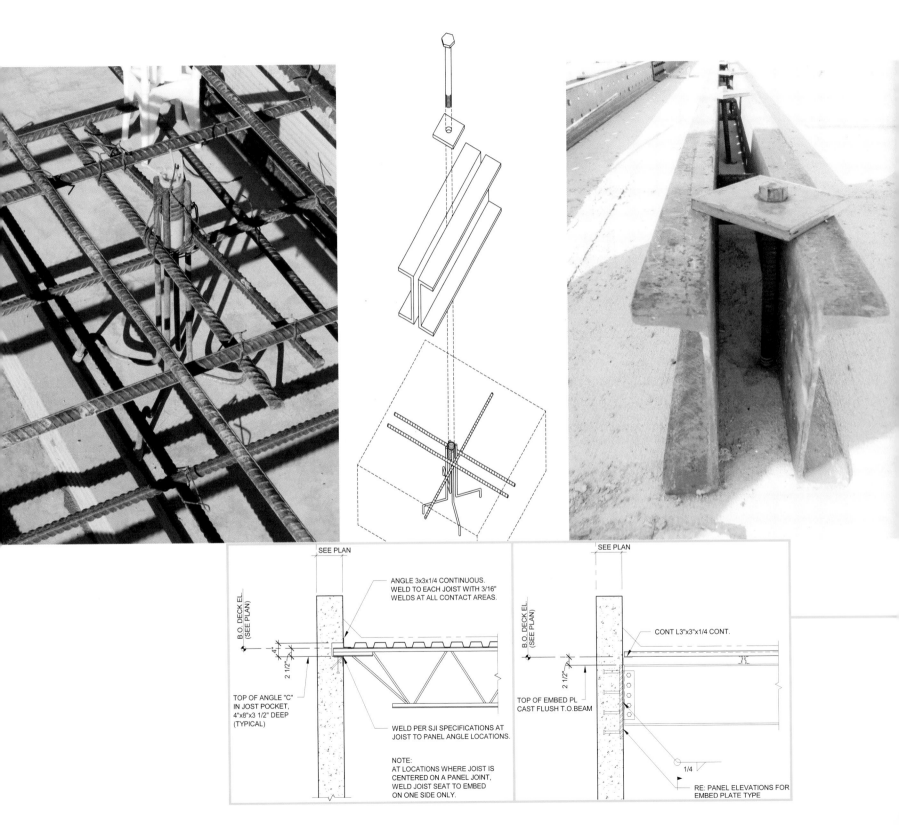

ANGLE 3x3x1/4 CONTINUOUS.
WELD TO EACH JOIST WITH 3/16"
WELDS AT ALL CONTACT AREAS.

SEE PLAN

B.O. DECK EL.
(SEE PLAN)

2 1/2"

TOP OF ANGLE "C"
IN JOST POCKET,
4"x8"x3 1/2" DEEP
(TYPICAL)

WELD PER SJI SPECIFICATIONS AT
JOIST TO PANEL ANGLE LOCATIONS.

NOTE:
AT LOCATIONS WHERE JOIST IS
CENTERED ON A PANEL JOINT,
WELD JOIST SEAT TO EMBED
ON ONE SIDE ONLY.

SEE PLAN

B.O. DECK EL.
(SEE PLAN)

2 1/2"

CONT L3"x3"x1/4 CONT.

TOP OF EMBED PL
CAST FLUSH T.O.BEAM

1/4

RE: PANEL ELEVATIONS FOR
EMBED PLATE TYPE

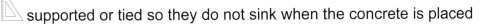 supported or tied so they do not sink when the concrete is placed

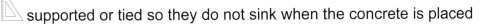 pointed in the correct direction (for example, lift inserts pointed to the top of panel)

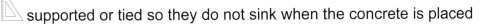 not set on formliners or reveals if possible

 for industrial products or where inserts are located in the interstitial space on commercial products, consider the use of **plastic inserts** for the patching of lifting inserts rather than patching mix

outside

tilt

inside

57

CONCRETE

$$C3S + H \longrightarrow C\text{-}S\text{-}H + CH$$

or

$$Ca3SiO5 + H2O \longrightarrow (CaO) \cdot (SiO2) \cdot (H2O)(gel) + Ca(OH)2$$

or

$$2Ca3SiO5 + 7H2O \longrightarrow 3(CaO) \cdot 2(SiO2) \cdot 4(H2O)(gel) + 3Ca(OH)2$$

$+$  $=$ CONCRETE

mix **delivered** to site generally **matches** mix **submitted** for approval (especially with respect to aggregate size)

ensure **aggregate size** is considered with the spacing of reinforcing so that **proper distribution** of aggregate can occur with the panel

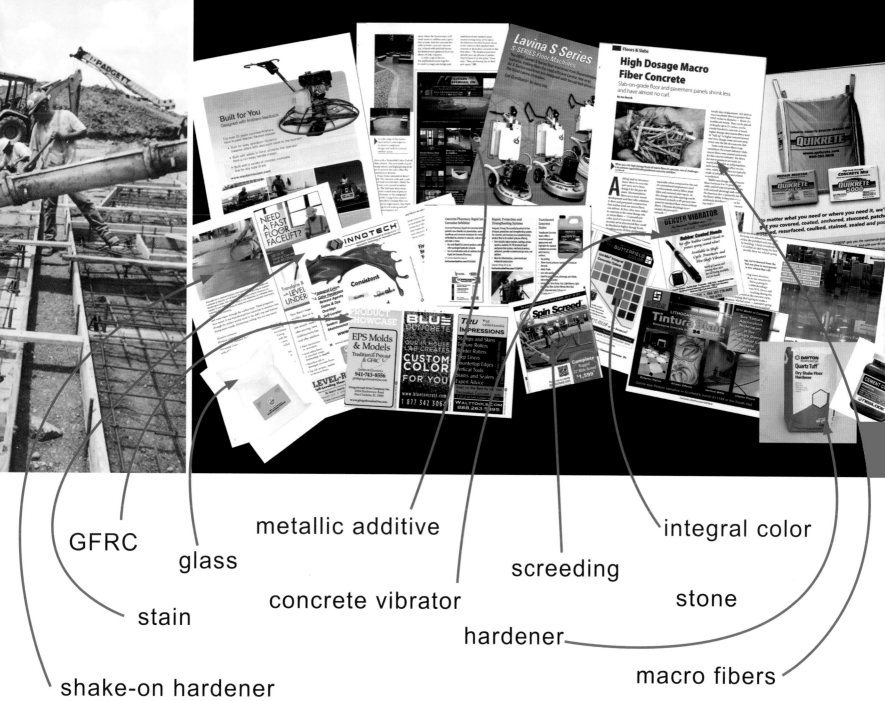

GFRC

glass

metallic additive

integral color

screeding

concrete vibrator

stone

stain

hardener

shake-on hardener

macro fibers

   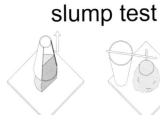

| fill cone 1/3 full | rod layer with 25 evenly spaced strokes | fill cone 2/3 full | rod layer with 25 evenly spaced strokes—not through the first layer | heap concrete on top of mold | rod layer with 25 evenly spaced strokes—not through the second layer | remove excess concrete from the top of cone | lift cone slowly and evenly | invert cone next to concrete, lay straight edge across top, and measure slump in inches |

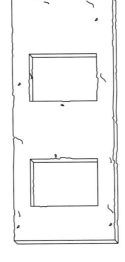 good vibration VS. bad vibration

△ **vibrated** as concrete is placed to **ensure consolidation** around embeds and reinforcement to **minimize voids**

⛑ **limit** the use of **fly ash** on panel mixes to ensure **quick curing** and clouding in the finish

△ technician **onsite testing** quality and taking **samples** for laboratory **testing**

△ **no cold** joints within panels

△ special care is taken for **cold- or hot-weather applications**

TILTWALL BEYOND

3"

KAWNEER TRIFAB 451-VG-101

SCHED GLAZING

451-HP-126 FLASHING CLIP

SCHED DIM

3" 3/4

KAWNEER 451T-HP-037 FLASHING CLIP SET IN CONTINOUS BED OF SEALANT

1" 1/4

SCHED DIM

SEALANT AND BACKER ROD-TYP

LINE OF FINISH OVERPOUR - TYP.

TILTWALL PANEL

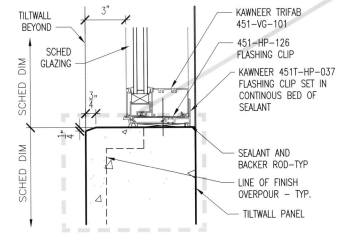

◯◯ applied **before casting:** integral color, formliner mold, exposed aggregate, precast overpour, thin brick

◯◯ applied **after casting:** broom, trowel, screed, float, stipple

◯◯ locally sourced / batch-plant sourced, thus varying characteristics per batch / delivery, resulting in an inconsistent natural finish

consider **aggregate type** for any **surface finish**—sand beds, "bush-hammering", etc

FINISH

ensure that when **patching** panels, proper care is given to the preparation of the area to be patched

onsite **stain** mock-up

hard-trowel finish at exposed portions of panel (for example, parapet) and around most doors and windows

 proper application of curing membrane

 the casting surface is the most important contributor to the level of finish on the panels

 prior to painting and patching, ensure that the panels are sufficiently clean so that necessary bonding can occur

◯◯ onsite **paint and stone** mock-up

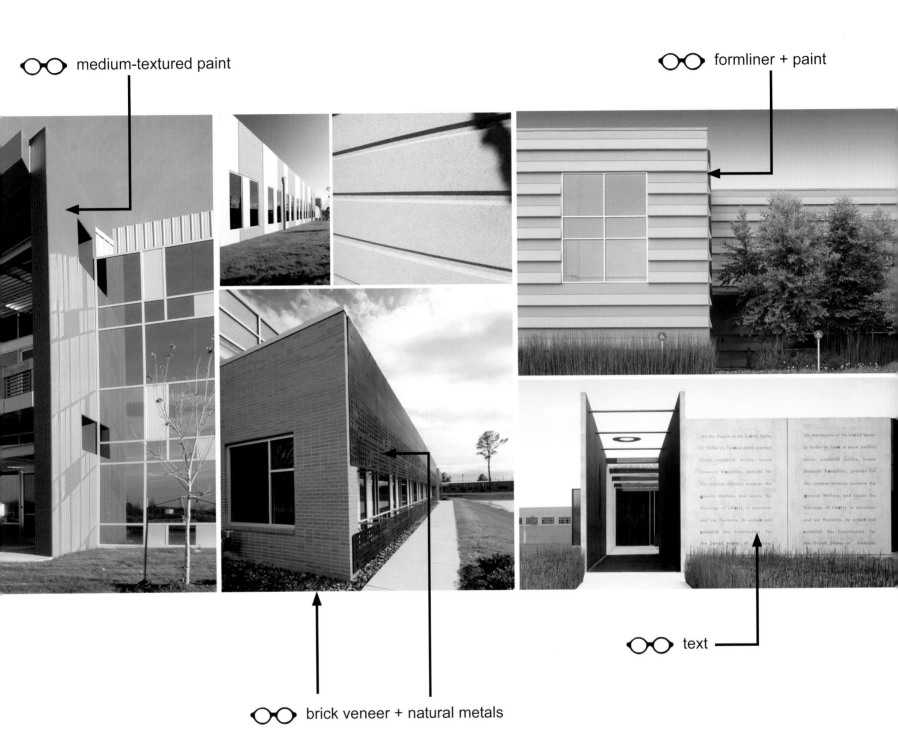

medium-textured paint

formliner + paint

brick veneer + natural metals

text

bush-hammered

stained

aluminum cladding

stone

THE LIFT

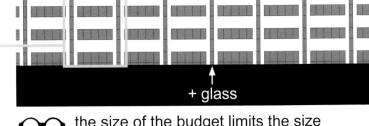

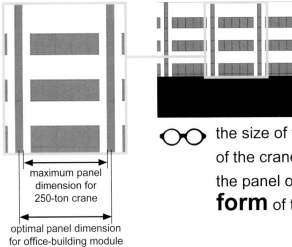

+ glass

maximum panel
dimension for
250-ton crane

optimal panel dimension
for office-building module

 the size of the budget limits the size
of the crane which limits the size of
the panel occasionally driving the
form of the building

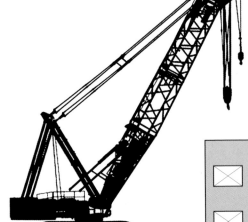

whether picking from the slab or from a crane-road, ensure that the integrity of the
lifting surface is measured and considered by the rigging engineer to ensure
appropriate load distribution

adequate **cribbing and blocking** to avoid damaging slab-on-ground or other structural elements

verify capacity at the maximum reach

the crane is the most expensive part of the panel erection process—manage the **panel size** as much as possible to limit the **crane size**

 lift inserts have been **located**, **cleaned** and **tested**

rigging cables match specified minimum lengths
on lifting design documents

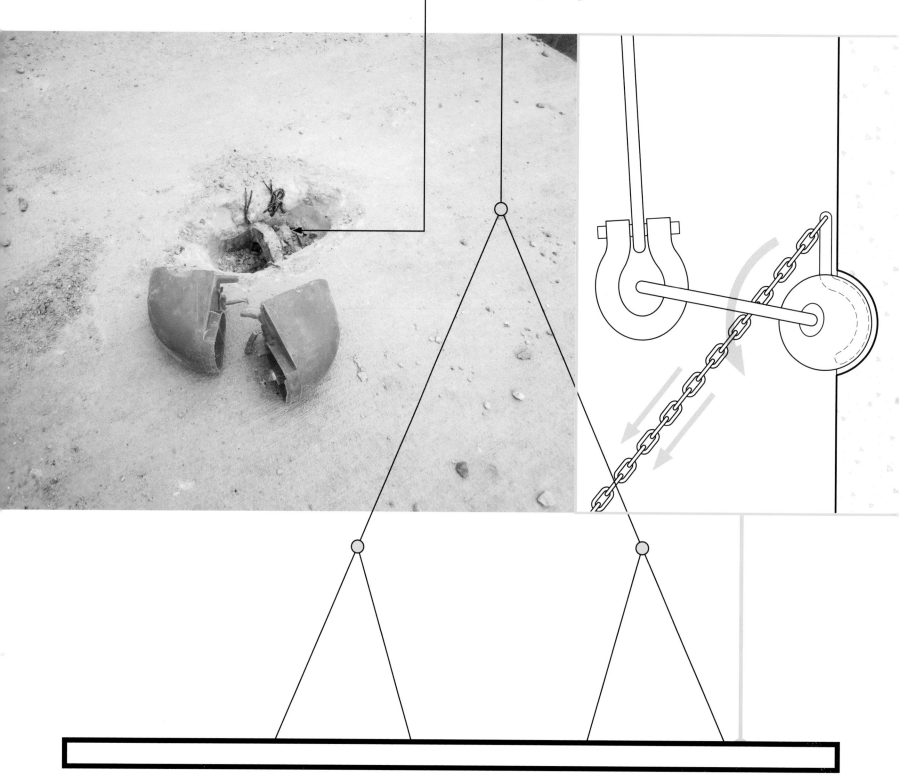

△ emergency lifting plates are available if lift insert cannot be found

△ shims and erection kick-plates are in the proper locations

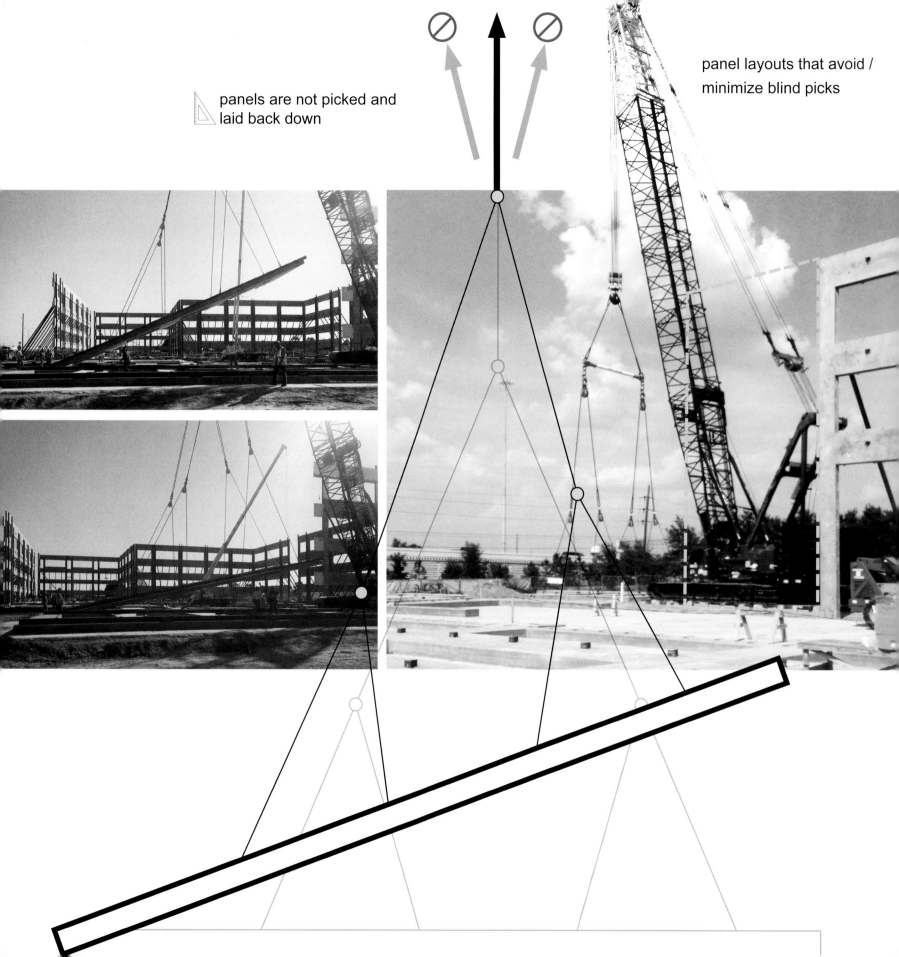

panels are not picked and
laid back down

panel layouts that avoid /
minimize blind picks

BRACING

for **multi-story** projects, it's always best for the overall schedule to **brace** the **outside**

72

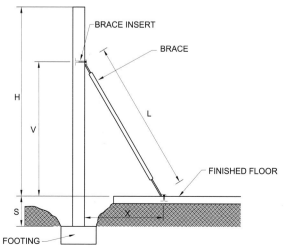

△ not used to crank the panels in or out during steel erection

△ **left in place** until a certain amount of the interior
structure is placed

BRACE INSERT

BRACE

H

V

L

FINISHED FLOOR

S

FOOTING

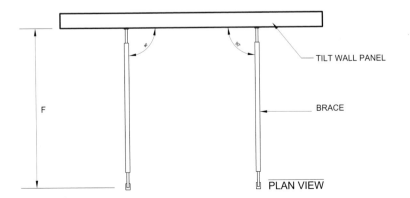

TILT WALL PANEL

F

BRACE

PLAN VIEW

 for **larger panels**, an assist **crane** is often required to handle bracing

 when bracing to the outside, consider the use of auger-style dead men to prevent the need to remove cast-in-place dead men (coordination of utilities, landscaping, etc)

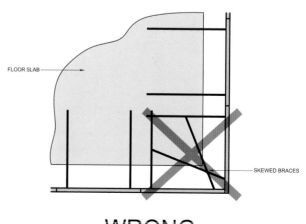

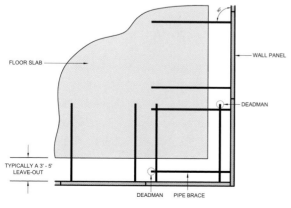

FLOOR SLAB

WALL PANEL

DEADMAN

TYPICALLY A 3' - 5'
LEAVE-OUT

DEADMAN   PIPE BRACE

RIGHT

FLOOR SLAB

SKEWED BRACES

WRONG

75

STACKING | SPLICING

siting of the building footprint must accomodate casting- and storage-area for future upper-level floors

← BUILDING
ENVELOPE

SLAB IS PREPARED
AND TILT-UP PANELS
ARE CAST ON SLAB
AND CASTING BEDS
ONSITE

TILT-UP PANELS ARE
SET AND BRACED
TO SLAB OR TO
EXTERIOR

STRUCTURAL
FRAMING IS TIED IN
TO TILT-UP PANELS

kick plates and temporary connections have been prepped and are ready for for panel setting

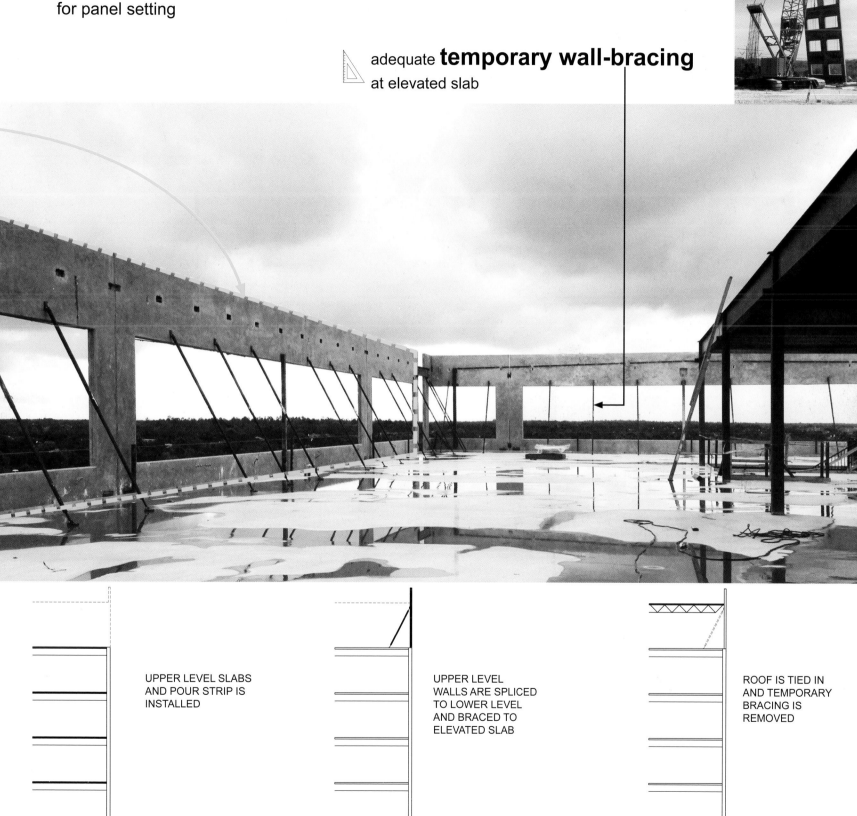

adequate **temporary wall-bracing**
at elevated slab

UPPER LEVEL SLABS AND POUR STRIP IS INSTALLED

UPPER LEVEL WALLS ARE SPLICED TO LOWER LEVEL AND BRACED TO ELEVATED SLAB

ROOF IS TIED IN AND TEMPORARY BRACING IS REMOVED

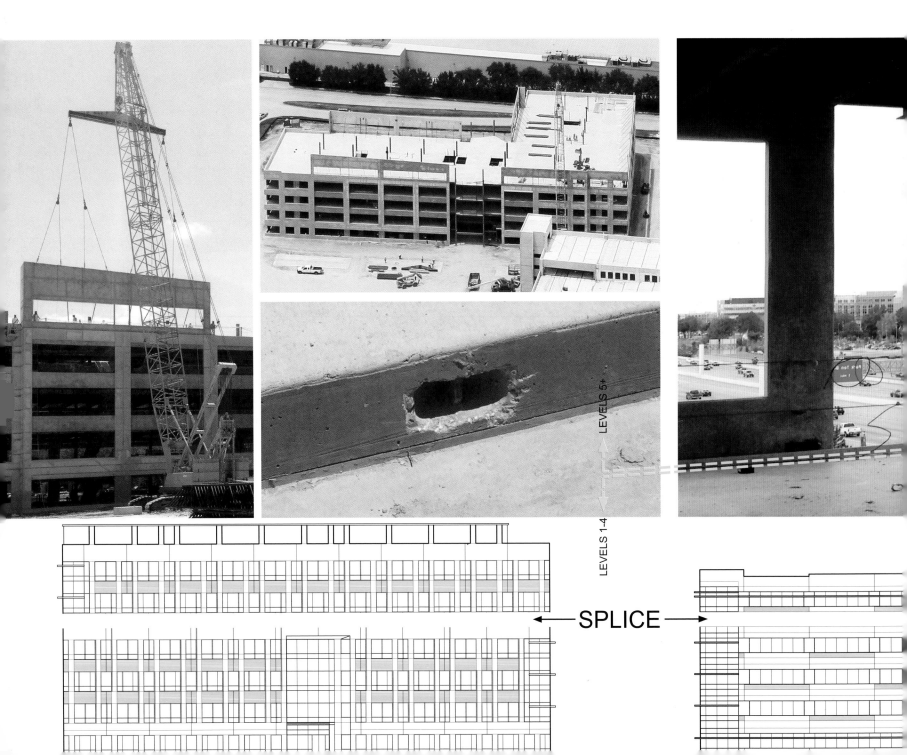

crane boom and counterweights as well as lifted panel do not hit panels already erected

LEVELS 5+

LEVELS 1-4

← SPLICE →

LEVELS 5+

LEVELS 1-4

INTERIOR

EXTERIOR

"Buildings don't come into being without a client and his or her aspirations, or without market appraisal, without a budget and the whole tedious steps of competitive tendering, value engineering and all the rest. Because construction is so deeply involved with the market and its various mechanisms it might be naïve to expect an architecture "critical" of the conditions of its emergence within the scope of an award dedicated exclusively to built buildings. It might be foolish to still expect works that question the status quo of our culture such as the all engrossing presence of commercialism and the culture of the spectacle, or that address transparently the ubiquitous pre-condition of a mediated awareness with its own economy of attention. But then again—it might be the only appropriate thing to do."

"Is There Such a Thing As Critical Architecture?",
Sauerbach Hutton Archive

Reconciling trends[i]

The Preface of this book starts with a notion of two guiding forces, the "market" and "latent theoretical content" or potential. Chapter 5 will deal with the latter, while this section will explore all things market related. This assumes that Tilt Wall is already in the market, forms a market or otherwise is established as a force to be reckoned with in the architectural market. Here, two seemingly independent bypassing messages have been communicated. One is an encouragement for architects to participate in this market through engagement and research—which implies that they are not. The second argument is that it has become a market by its use as a prevalent method of building by non-architects, not just stateside but globally. This suggests that many in the construction and development industry have embraced and begun to explore Tilt Wall's potential. Reconciling these two messages involves describing the contemporary state of Tilt Wall construction after its promising history. As observed at the end of Chapter 1, the current condition is one of lack of interest by architects, mirrored by rapid acceptance and deployment by clients, in particular developers and contractors, who make up a great deal of the everyday practice environment. Contractors are here included as "clients" given the trends in procurement and project delivery towards Design Build and Construction Manager at Risk systems in which the architect actually works for the contractor. Client adaptation is, for architects, a kind of mainstream condition and is not intended to leave aside the capital "A" architectural potential of the Tilt Wall method explored in later chapters. Indeed it is reconciling Tilt Wall's mainstream dilemma that is the foundation of its real potential.

Why do contractors take such great interest in it?

Contractors capable of and with an expertise in this way of building have proliferated in the last 20 years or so. Over 10,000 buildings, enclosing more than 650 million square feet, are constructed each year using this construction method,[ii] with hundreds of firms claiming experience in this methodology. It has spread to every state in the U.S., into Canada and the balance of the Americas. Many major U.S. corporations travel this method to Mexico and Central / South America for their manufacturing and distribution facilities. In addition it has gained legitimacy in New

Zealand, Australia, Russia and the Dominican Republic, with many European countries and the Middle East quickly adopting the method as well. Concrete is a fundamental material in building and Tilt Wall relies upon it, thus there is a natural fit between skills every contractor has to broker and the technique. Consequently, the learning curve for Tilt Wall as opposed to titanium cladding, for example, is favorable. There has also been a move towards design-build delivery methods in which the cost of the project has been separated from its design; contractors lead this delivery specialization and the architect is contracted to them. Under this delivery methodology, with cost and time as a driving factor, the architect's formal propositions are often predetermined by construction methodology. Ultimately contractors like Tilt Wall because they can control it to a degree they cannot in other methodologies particularly through self-performance. Four major forces are at work in this control: management simplicity, trade reduction, reduced-risk reward on profit and safety, and ultimately schedule flexibility. As Tilt Wall is based on simple construction techniques and details, it can be managed with smaller, more junior staff assignments and capability. Second, as the perimeter structure and the majority of the façade fall under one trade, it requires less coordination. There are no lead times for brick trades or integral cladding systems on the critical path—the skin is the structure no matter what it is or is not clad with. Third, less trades and simplified construction techniques managed by fewer personnel results in a better return on staff with reduced risk in terms of injury, and in profitability. Finally, because the work can be coordinated and performed without significant lead time or up-front coordination before the start of construction, it is good work for contractors to use in gap times or to maintain crucial staff balance, simply put many see it as a "quick hit".[iii]

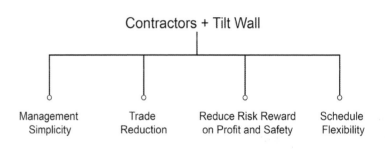

While there can be a balance between cost and content, architects have preferred to ignore the possibility in favor of formal propositions. This considered at the mainstream level of design practice and building (shall we say a market-based level) runs counter to trends that indicate that the first entity many potential clients turn to when considering building is a contractor rather than an architect.[iv] The World of Concrete convention, at over 65,000 annual participants, is one of if not the biggest convention in Las Vegas. It is attended almost entirely by non-architects like contractors, sub-contractors, vendors and specialty equipment manufacturers. Importantly many developers, facility managers and corporate clients attend this event. The professional trade organization, the Tilt-Up Concrete

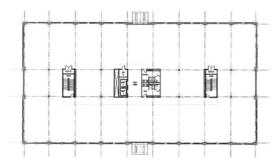

Conventional Construction
A typical office product based on a 25,300 square foot floor-plate and constructed conventionally has …

- perimeter columns
- columns in lease areas
- 43'–0" deep lease area along perimeter
- total columns used = 40

Based on a preliminary pricing exercise completed in May 2007, this floor-plate constructed at four stories would cost approximately $10.9 million.

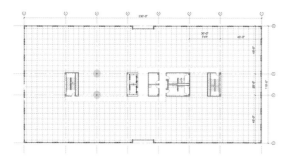

Tilt Wall Construction
By comparison, the same 25,300 square foot floor-plate built utilizing tilt wall construction has …

- no columns at the building's perimeter
- 5' leasing grid
- 45'–0" column free lease space along perimeter
- centrally located data / electrical rooms
- total columns used = 12

Based on a preliminary pricing exercise, this floor-plate constructed at 4 stories would cost approximately $8.8 million …

- a savings of $2.1 million over conventional construction
- a savings of 10% in steel tonnage

Association holds its annual meeting and awards ceremony at this convention, exhibiting a perfect intersection of material and technique at the scale of the entire architectural, engineering, and construction industry (practically sans architects). The TCA, founded in 1986 and funded by major industry stalwarts such as the Portland Cement Association, the Concrete Reinforcing Steel Institute and the National Ready Mixed Concrete Association, has over 1,000 members and is growing globally. Thus the material they use in nearly every project, the vendors that sell the equipment to manipulate it and their clients all converge. While Tilt Wall is not the main method on display at the convention, having to compete with precast and unitized concrete systems, it is in the mix and that is what is important to its success as a rising method with contractors.

Why do developers take such great interest in it?

Developers survive at the nexus of financing and marketable form. Recent economic challenges have begun to distort the equation even further in favor of cost-driven form and Tilt Wall delivers on that desire. Traditionally many avant-garde architectural practices have eschewed the developer as a client, while paradoxically many of the profession's leading mainstream firms rely upon them as base clients. In the commoditized world of development architecture, which makes up a great majority of the built environment, Tilt Wall is fast becoming a product of preference. On certain building types, office and corporate projects in particular, it will perform at averages of $10 per square foot, less expensive than comparably sized and architecturally appointed conventional construction. The result has been a rapid adaptation of this "product", this way of building a comparable asset for less. From a developer's point of view, speed of delivery with a high-quality product is extremely desirable. If the standard for a traditionally defined Class A in a market is high-rise, then the goal is to replicate (or improve upon) those standards with a Tilt Wall method. Expansive vision glass, high-quality finish materials, and structured parking can be equaled in the technique. In addition, the efficiency of the space can be optimized due to larger floor plates being cost effective, but this can produce more interior floor, which does not appeal to smaller tenants who cannot use that space as effectively. The major challenge is balancing aesthetics with efficiency, what many developers call "sex appeal". As Tilt Wall buildings

aren't high-rise, invention with the method is required to bridge that exterior design gap for all but height, delivered at a material occupancy cost discount.

But it is not cost or market forces alone that have attracted many of the nation's most prominent developers to include Tilt Wall building in their portfolios—it is also the investment-grade level of quality that the method can achieve. Yet again, it is non-architects that have recognized (either first or more frequently, or both) the upstream potential of the method. Developers have been challenging architects to explore more and with different building types using Tilt Wall, and have been demanding exacting standards from the contractors building them. This has resulted in financing sources on the equity side and banking / financing side to begin to accept Tilt Wall construction as a Class A methodology. Tilt Wall is now accepted by many of the most prominent corporations, who are signing leases for 10 or more years. The developers that decided to operate with Tilt Wall construction have opportunities to generate higher returns for their clients because the spreads between total costs and exit prices have widened. Those prices are ultimately indexed to capitalization rates. From Wikipedia:

> In real estate investment, real property is often valued according to projected capitalization rates used as investment criteria. This is done by algebraic manipulation of the formula below:
>
> Capital Cost (asset price) = Net Operating Income / Capitalization Rate
>
> Capitalization rates, or cap rates, provide a tool for investors to use for roughly valuing a property based on its Net Operating Income. For example, if a real estate investment provides $160,000 a year in Net Operating Income and similar properties have sold based on 8% cap rates, the subject property can be roughly valued at $2,000,000 because $160,000 divided by 8% (0.08) equals 2,000,000. A comparatively lower cap rate for a property would indicate less risk associated with the investment (increasing demand for the product), and a comparatively higher cap rate for a property might indicate more risk (reduced demand for the product).

$$\text{Capitalization Rate} = \frac{\text{annual net operating income}}{\text{cost (or value)}}$$

According to an informal survey among developers taken in mid 2013, recent transactions for Tilt Wall office buildings will now be trading for less than a 7 percent cap, which is on par with traditional construction methods.[v]

Tilt Wall construction and its architectural potential meet at the nexus of the years' long ongoing economic transformation and developers have capitalized on that. They address the dilemma of how to produce meaningful architectural form in an economical manner. Developers are already transforming their approach with design-build and negotiated delivery—they control and dictate construction methods and Tilt Wall is becoming a prevalent option. The federal government now mimics the private sector and issues design-build lease back projects, often more frequently than traditional architect / engineer led procurements. They now allow Tilt Wall construction in most solicitation for offer documents. Many state and local public institutions have followed suit. This has all been brought on by the desire to capture the efficiency of the private market, established and led by developers and enabled in part by Tilt Wall construction.

Why architects should take an interest in it

Why they have not is covered elsewhere in this book, particularly in Chapter 5. The argument thus far assumes that architects have understood this method as the domain of the contractor, thus constituting a lower order of making not robust enough for manifesting "meaningful form". In so doing architects experience a gap in the repertoire of design approaches capitalized upon by "the market", defined as clients and constituent allies in their business.

At a time of heightened interest in re-engaging the actual *making* of buildings, Tilt Wall construction is a trend architects must be on the forefront of regarding formal innovation. This is an important notion given the history of "missing the boat" on two current areas of intense architectural industry focus: sustainability and Building Information Modeling (BIM). As much as these issues currently consume the space of popular architectural effort and dialog, it is frequently argued and commonly understood that architecture as a discipline put itself behind the curve in both areas. The formation of the U.S. Green Building Council (USGBC), and the

implementation of BIM as an industry standard were both initiated outside of the discipline or worse, in spite of it.[vi] This need not happen with the formal architectural engagement of economical methods of building such as Tilt Wall. Particularly given the methods formal, structural, sustainable, and global potential attributes; all concerns traditionally in the bandwidth of architecture's leading trends.

Possibilities of tilt wall

Formal

Tilt Wall construction is a method of site casting walls for a building utilizing the building's slab as a casting / forming surface. While there are many variations and adaptive solutions to specific site conditions, the premise is the same. The shape and profile of the panels are extremely varied including panels that are convex, concave, "L" shaped, "F" shaped, and "E" shaped. Flag panels, rhombus shapes, puzzle configurations and round panels illustrate just a few cases. The size, shape, and general configuration of panels is ultimately mutable which aligns with the current interest in digitally generated forms and patterns. Combined with finish options (mentioned above), the formal repertoire available with this technique has only just begun to be exploited. The walls are load bearing; in effect they become planar columns. The panels are not limited to sizes that can be trucked to the site, nor are they limited to being only very large. They can be strategized in the same manner as pre-cast panels, with spandrel pieces that rest between panels and not on structural foundations; they can also have textures added and be finished in nearly every material or process. The major advantage they have over pre-cast is speed. On many building types the entire process of utilizing the method saves between six and eight weeks of time.

Illustrating some aspects of the formal possibilities of the method, the TCA website notes the following statistics:

Target Import Warehouse
Savannah, GA

Gateway Theatre of Shopping
Umhlunga Rocks, South Africa

El Paso County Detention Facility
El Paso, TX

The Gap/Old Navy Distribution Center
Fishkill, NY

Lucky Street Garage
Hollywood, FL

Ocean Center
Daytona Beach, FL

The Toho Water Authority
Kissimmee, FL

FAU Innovation Village
Boca Raton, FL

Voyager Academy Charter School
Durham, NC

The largest footprint is 2,029,554 sq ft (188,546 sq m)

The largest total floor area is 3,420,000 sq ft (317,718 sq m)

The largest total wall area is 1,400,000 sq ft (130,060 sq m)

The largest panel is 2,950 sq ft (274.1 sq m)

The most panels in a building is 1,310

The tallest panel is 96 ft 9 in (29.49 m)

The heaviest panel is 339,000 lbs (153,768 kg)

The widest panel is 81 ft 7.5 in (24.88 m)

The tallest cantilever wall is 70 ft 6 in (21.49 m)

Structural

Simply stated Tilt Wall's structural possibilities are as open-ended as they are unexplored. Tilt Wall is fundamentally structural; while many slab or site-cast concrete panels were employed as cladding, in its most effective format, Tilt Wall is a load-bearing wall method of building. Once the exterior shell is complete the structure is erected internally and tied to the exterior panels forming a complete vertical and lateral load-resisting system. The walls carry the vertical loads, can act to resist shear and torque forces, and are both expression and structure. The engineering itself requires some particular expertise but can be performed by any structural consultant supplied with the extant and replete resources available. The basics include the design of reinforcing bars, the concrete coverage of those bars, and the resultant calculation of the panel thickness based upon maximum dimensions, overall weight, and the requirements for vertical and lateral loads. The forces in effect during lifting and bracing points for the panels, and the design of the footing and main structure connections are also slightly different than similar issues for conventional construction.

When used as load-bearing elements, the panels support the dead and live load from all levels of framing, whether floor or roof. This results in a downward, axial force within the panel. As the panel deflects out-of-plane, this axial force induces additional bending loads a structural consultant must consider to not only reinforce the panel properly, but also assure the resulting deflection meets both code and aesthetic concerns. Simultaneous with the vertical loads, the Tilt Wall experiences pressure on its face due to wind, or an idealized application of the inertial forces created by an

earthquake. When all of the loads are considered in concert, the panel is analyzed for flexure and compressive force. It is noted that today's codes prescribe wind and seismic forces that are expected to be exceeded only once or twice in a 50-year lifespan.

For single-story buildings, and two-story buildings with large floor-to-floor heights, the Tilt Wall design is primarily controlled by the out-of-plane bending resulting from the wind or an earthquake event. Forces are distributed evenly across the width of the panel unless an opening exists. In this case, the total load on the Tilt Wall is concentrated in the remaining panel legs (or jambs) that span continuous between out-of-plane horizontal supports (from foundation to roof, for example). In multi-story construction higher than two stories, the panel design is primarily controlled by the compressive forces in the legs, resulting from the accumulation of dead and live loads of the floors and roof above. The highest flexural forces in these cases will most likely be from the lifting operation, but there are several techniques developed by the industry to handle these conditions beyond adjusting the panel thickness, opening configuration, or reinforcement.

In nearly all buildings, regardless of the number of stories, the panels are used as concrete shear walls to make the building laterally stable, or to assist other lateral load-resisting elements. Floor and roof framing is generally designed as a structural diaphragm that collects the lateral loads from the perpendicular face of the building and distributes them to the Tilt Walls running in the parallel direction as the force being applied. In addition to the vertical and out-of-plane loads discussed above, structural consultants take into account both the connection of these diaphragms to the panels and the resulting shear in the plane of the Tilt Wall to specify the appropriate thickness and reinforcement. Unlike previous perceptions about standard practice for pre-cast type panels, Tilt Walls are generally only connected together when required for stability, or to ensure there is no differential movement when an opening crosses a panel joint.

Again, none of the engineering design work is outside of common industry knowledge. The method has evolved to exceed its early rule of thumb estimates of efficiency regarding the area of slab to wall. The engineering of a panel requires some degree of particular expertise and a familiarity with the design principles that apply to the field of tilt-up design. The structural engineer is charged with determining the thickness and strength

Blast Resistance Theory

Medium Level of Protection
Low Threats per UFC 4-010-02 (Jan 2007)
Conventional Construction Standoff
148' to Perimeter
82' to Internal Parking

Building Damage
Minor damage—building damage will be economically repairable. Space in and around damaged area can be used and will be fully functional after cleanup and repairs.

Door and Glazing Hazards
Glazing will fracture, remain in the frame and result in a minimum hazard consisting of glass dust and slivers. Doors will stay in frames, but will not be reusable.

Potential Injury
Personnel in damage area potentially suffer minor to moderate injuries, but fatalities are unlikely. Personnel in areas outside damaged areas will potentially experience superficial injuries.

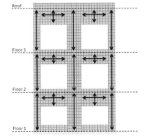

Progressive Collapse Theory

Progressive Collapse is defined in the commentary of the American Society of Civil Engineers Standard 7 **Minimum Design Loads for Buildings and Other Structures** (ASCE 7) as:

The spread of an initial local failure from element to element, eventually resulting in the collapse of an entire structure or a disproportionately large part of it.

Sustain local damage with the structural system as a whole remaining stable and not being damaged to an extent disproportionate to the original local damage.

Structures are designed to **limit the effects of local collapse** and to prevent or minimize progressive collapse.

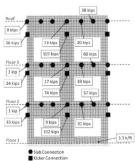

Structural Modifications

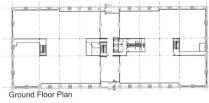

Ground Floor Plan

Typical Floor Plan

Changes to the floor layout were not necessary in order to comply with progressive collapse and blast resistance. The major structural modification was the addition of interstitial footings.

Another option is a continuous grade beam along the entire perimeter of the building to support the intermediate panel legs of adjacent 30-foot-wide Tilt Wall panels for progressive-collapse resistance.

Structural Modifications @ Panels

Kickers are installed between the floor / roof systems to the bottom of the spandrel panel to provide lateral restraint for progressive collapse resistance. Connections between the Tilt Wall panels and the floor diaphragms are upgraded for blast-loading and progressive-collapse resistance.

Typical Panel Elevation

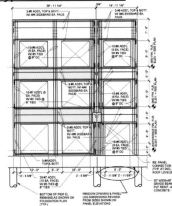

Spandrel panels at each level of the building are designed to span the full width of the typical 30 foot wide Tilt Wall panel, supporting the gravity loads of one floor and one story of the Tilt Wall panel, for progressive-collapse resistance.

Tilt Wall panel legs are reinforced to support the load from adjacent panels for progressive-collapse resistance.

---

of the concrete, as well as the proportions of internal reinforcing steel that will be required to satisfy the conditions dictated by the given wall panel sizes, spans, and imposed loadings. There will generally be numerous unique cases on any single building that will need to be investigated and designed. Wall panel elements that are solid or nearly solid surfaces will have different structural thicknesses and reinforcing arrangements than neighboring panel elements which might have large surface openings (doors and windows), a higher intensity of applied gravity loads, or are subjected to larger winds and seismic forces due to their relative size and position within the building. A structural engineer with sufficient experience designing tilt-up projects can recognize and identify potential problem areas with a proposed building plan layout, or proposed architectural façade design scheme, and help formulate acceptable solutions.[vii]

Panels as low as is conceivable and as tall as 97 feet have been executed. Building size is unlimited with single projects approaching over three million square feet on the ground. Simultaneously, many small religious and residential projects have made the popular architectural press. Beyond buildings, Tilt Wall has been used in non-building structures from anything as basic as utility boxes to more prominent monuments like the Korean War Memorial in Kansas City. Tilt Wall is capable of multi-story construction in both its load bearing capacity and as cladding and can be used to cast onsite structural elements for use in stadium or other infrastructural architecture. In all cases, the concrete structures are long lasting and, when properly designed, capable of numerous structural response to natural disasters, terrorism threats and longevity standards.

Tilt Wall has been used to complete structures capable of simultaneously resisting a Category 5 hurricane and F3 tornado—just such a structure remained undamaged in the Houston area after Hurricane Rita. More recently Tilt Wall structures have now been completed that meet or exceed the most up to date issuance of the Unified Facilities Criteria (UFC) for Department of Defense Antiterrorism Standards: the UFC 4-023-03 (July 2010), the Occupancy Category III per UFC 3-301-01 (Jan 2010) which relates to progressive collapse resistance mandated after collapse of the Murrah Federal Building in Oklahoma City. In addition, for blast-resistance analysis at the medium level of protection buildings are being designed to meet threats per UFC 4-010-01 (Jan 2007-2010).

Sustainability

Sustainability, green-building and environmental considerations are well understood and well documented in other resources. Tilt Wall construction, like any system, technique or component of the construction process, is only a piece of a larger set of considerations. For its part, Tilt Wall intersects sustainability in terms of its constituent material concrete, its comprehension as a system and its general characteristics regarding sustainable best-practice goals.

In so far as concrete is a sustainable construction medium, so is Tilt Wall. It supplies similar benefits as in any concrete system including the use of fly ash additives that reduce industrial byproduct land fill, as well as slag, silica fume, foundry sand and lime sludge. Once mixed into concrete they are not only harmless they are also recyclable. Concrete is a naturally non off-gassing material and thus low VOC. Certain add mixtures can allow for naturally finished concrete surfaces to absorb carbon dioxide from the atmosphere, as well as provide for self-cleaning organics that may have been oxidized resulting in contributions to air quality in and around the structure. Addressed as a system, some important things are elaborated elsewhere such as the site-driven nature of the method and its low carbon footprint. Augmenting that is the actual product of load-bearing wall construction, or the building itself. The nature of Tilt Wall design accommodates—more so than nearly any other method of building—the ability to minimize exterior joints and thus concomitant leaks and moisture penetration. Joints are minimized to those between wall and roof, around penetrations and most important in terms of efficiency, between panels. This reduction of joints retards air infiltration and induces better overall building performance. When Tilt Wall panels are designed as sandwich panels with rigid insulation, they offer a very high exterior wall "R" value, contributing even more to building quality. With or without integrated insulation, the natural byproduct of cast concrete walls is thermal massing, again a sustainable best-practice in many climates. While there are many other important aspects of Tilt Wall that contribute to sustainability, these are inherent in the system itself and, in combination, unique to the method.

Structural Modifications @ Roof

Partial Roof Framing *Typical 30' x 45' Bay*

The roof framing system was upgraded to steel beams in lieu of steel joists.

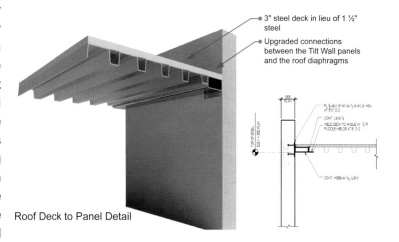

- 3" steel deck in lieu of 1 ½" steel
- Upgraded connections between the Tilt Wall panels and the roof diaphragms

Roof Deck to Panel Detail

Panel Modifications

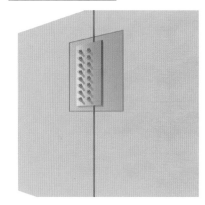

Typical Panel Connection
Steel plates are provided on all Tilt Wall panel joints at each level of the building to transfer load to the adjacent panels for progressive-collapse resistance.

Via more general consideration, Tilt Wall is a site-performed method and, unlike precast construction, requires no truck trips to move the finished product to site, thus reducing embodied energy and fossil fuel consumption. Resource planning can result in the higher use of local materials for forming and working the concrete, as well as the potential reduction of land fill trips and waste.

Global

As stated elsewhere, in some ways Tilt Wall construction has more in common with parochial methods of building such as "barn raising", at least as a commercial application of the technique rather than as any kind of communal activity. Much of the skill of Tilt Wall design lies in the coordination of engineering and architecture. Designing and calculating the correct reinforcing for the panels requires an experienced engineer. While many regions that could benefit from this way of building may not have such talent on site, this is true for competing Western methods of construction as well. The recent market globalization and preference for American know-how in emerging markets such as China and the Middle East have taught us that talent travels and locals learn quickly.

Regarding physical labor, the problem is less complex. The layout and construction of forms is a byproduct of rough carpentry and basic skills. Once designed, the process of lifting and placing the panels is a technique that can be performed by the lowest common denominator labor force. Training is also available and accessible at most levels of construction experience. Tilt Wall construction has the capacity to be inclusive and engaged with local culture and labor, in emerging markets and, potentially, as a disaster mitigation technology. Non-government and relief organization programs such as Architecture for Humanity's "Design Like You Give a Damn" point this out as key to progress and recovery.

Mainstream, everyday, and market

Much of Tilt Wall's popularity on the building front and its neglect in the architectural profession is captured by the repeated use of the term and concept "mainstream". By "mainstream", sometimes interchangeable with "everyday", both architecture and the market are being addressed or, in a way, indexed to one another. The mainstream market concept as comprised of developers and contractors attempts to capture the force behind the proliferation of Tilt Wall construction. It purposely means to

exclude the awe-inspiring Dubai skyline with its acrobatic high-rise towers and the redevelopment of New York post 9-11 (and so on) in favor of buildings that seem to escape critical investigation but occupy most of the built environment: the office parks, industrial parks and low-rise housing developments along with high- and low-end retail. Or corporate campus projects and—given the design build lease back procurement method employed by the GSA—even government buildings. All of these everyday project types far exceed those at the margins of mediocrity or excellence, or budget.

Thus, for mainstream architectural practice a simple indexing resolves the definition; the great majority of firms do this type of work. Only a small number of architectural practices are published in what is left of the print media nowadays. The subject seems to be verboten and is rarely openly discussed, and very few general practice architecture firms can afford to "be critical", experimental or avant-garde. That is not an argument for two tiers of practice as much as it is an attempt at "consciousness raising" for the normative bulk of practice, the mainstream. There are several, (outlined in this book, in Excursus 3) leading-edge practitioners who have explored the Tilt Wall medium. But their efforts are overwhelmed by the sheer number of projects built this way. Far too many of them undertaken without architects involved, or by architects who did not see the opportunity at hand. Tilt Wall is an economically accessible way to facilitate the experimental in general practice.

A note regarding market authenticity

It was implied in the Preface that Tilt Wall construction is a development dangerous for architecture to remain disengaged with. While it has been almost completely ignored in academia and vanguard practice, its prevalence and ubiquity has become alarming. In a city the size of Houston, 5 million square feet of industrial space alone is "absorbed" by the market annually,[viii] a majority of which is new construction built utilizing Tilt Wall methodology. In this category alone, each of the approximately 350 firms in the city could execute one 17,000 square foot project built this way each year. Tilt Wall construction has begun to jump building types at an amazing rate. Office buildings, educational facilities and government and institutional clients are utilizing the economies of Tilt Wall brought about by the speed of erection and the low cost of labor. Its structural flexibility is also being exploited and utilized; to meet the Department of Defense

93

As a fully integrated approach to building, Tilt Wall differs from its near cousin precast, and from pre-engineered / pre-manufactured systems, in that it is both structure and cladding at once. With Tilt Wall function and meaning are inextricably united, combining Semper's division of the symbolic and technical aspects of the content of vertical surfaces. In addition it differs from what constitutes the prevalent subjects of research in architecture because it operates at the scale of building rather than at the scale of components. It is a manner of onsite fabrication—one of the most frequent subjects of architectural investigation in any casual review of current media coverage—but it is not an emergent or digitally based Boolean technology; rather it is vestigial in a way. Evolved in terms of execution but fundamentally unchanged in its principals of execution.

A recent revived interest in pre-manufactured / pre-fabricated building components is evidenced by a plethora of popular style-related books recently published. They are dominant to the extent they sit next to the ubiquitous Frank Lloyd Wright books in the few remaining retail book stores. This appears, at first glance, to run counter to the strategy of new materials and forms for every project every time. It has however largely limited its scope to that small scale spectrum of designer houses Prefab as it is called by many, is a movement that combines the aforementioned modernist interest in mass production with the customized use of standard parts (utilized by Charles and Ray Eames in their eponymous house). Other inquiries into mass customization and "transfer technologies" have excited the craft-inclined DNA of designers and overcome building type and scale limitations (to some extent) for commissions at the top of the project-budget spectrum. Just see the five-year cycle of the rediscovery of shipping container architecture that has been going on since the early 80s. Yet this line of investigation seems to cloak the modern argument of mass production with a different type of production, neither of which is as site specific as Tilt Wall construction. For all of the study of the car industry, it is set up to make tens of thousands of repeated elements after the perfection of the processes are manifested in a prototype. A building is by nature a prototype every time.

and General Services Administration's blast resistance and progressive-collapse criteria, for example. These trends are mirrored nationally and its potential in second- and third-world economies promises to eclipse its current success. Countries such as Australia and South Africa now have chapters of the Tilt Wall Concrete Association, educating local contractors in the method.

A great deal of the everyday building activity is borne out of the fluctuating relationship between architecture and capitalism. After the stateside transmutation of European Modernism into "corporate" architecture in the late 1950s, as both a type of practice and type of building, critical practice soon created strategies of resistance to Taylorized corporate productions. Modernism began to be scrutinized as the legitimate means of assuring the relevance of architecture's social potential, offering alternative realities to the purported suppression of individualism, and the de-humanizing effects inherent in the mixing of culture and economics.[x] Architecture, modernism and capitalism continue to be central characters in the developing plot of what constitutes legitimate meaning in design, as explored in the next chapter, and Excursus 2.

Meaningful architectural operations are possible in concert with capitalist society, and are as important to the mundane as they are to the elite. Many Tilt Wall projects constitute a majority of the physical environment, but resist the type of critical investigation that leading practice implies. They are projects driven by capital flow, financing, real estate formulas and expediency of construction so that any perceived non-essential speculation is expunged from the timeline. In the 21st century, architecture remains a "slow and naked discipline".[xi] Projects regularly exceed two years from inception to completion, making the software they were documented in no longer relevant to the project. Tilt Wall construction's low-tech method yields a high speed of erection that has reduced this time to under six months in some cases, allowing projects the chance to "lock in" favorable pricing and avoid the uncertainty of future escalation. Should the low budget / small business clients, who employ the bulk of workers in the country, be excluded from the benefits of high design because architecture can only happen when money is not a factor?

There has been much written of late in the above described debate of capitalism, market forces and their legitimate relationship to the process of design and ultimate architectural expression. Koolhaas and his followers have been circumscribed by Kwinter; accusations of pragmatism displaces the impossibility of planning in the context of the market with a pessimistic "rote numeric sequencing of market behaviors".[xii] This is similar perhaps to the propositions imputed to Reiser and Umemoto's work by Picon, in his description of their digitally derived architectural form as the expression of underlying fields of forces or "emergence". These positions may be counterbalanced by Tim Love's work and writing which represents a more practical approach to form making, based upon engaging the value propositions of developers as creative architectural material, rather than criteria to be eschewed. While neither of these arguments is easily applicable to the exploration of Tilt Wall construction, they do establish a framework by which to propose how the method can be investigated.

In the context of capital flow and economic forces on one side and nontraditional or non-patron client types such as developers on the other, it is not a stretch to begin to understand Tilt Wall's ubiquity as a new global vernacular. That is to say, a way of "making" that is indigenous to capital systems that require the fast and cheap, and which produce forms which transmit a value system and cultural makeup in much the same way a thatch roofed hut does in parts of Africa. The structure of the methodology indicates something of the cultural load and therefore the meaning, albeit not what traditionally constitutes the preferred content of architecture. The use of Tilt Wall is an enabler of economic expediency in construction, hence it's exponentially growing popularity. That is, in some ways, bad news. Optimistically, the method's architectural potential does not dilute its efficiency, nor its ultimate basic cost structure. It simply lacks investigation at the scale equivalent to the popularity of its use.

i   As in the Acknowledgements, much is owed to Messrs. Rogge, Griffin, Reed and Penrod on technical matters and Paul Coonrod and Carleton Riser for market / development input.

ii Tilt Up Concrete Association.

iii Kevin Rogge Harvey Cleary Builders (from an email exchange).

iv Our own statistics at Powers Brown Architecture show that well over 60 percent of our work has a first point of contact with a contractor.

v   Taken by the author, including: Transwestern Development, Stream Realty, Trammell Crow, Corporate Office Property Trust and Duke Development.

vi BIM really stems from contractor strategies to reduce intent lawsuits.

vii Load bearing tilt-up concrete wall panels are most often designed to span vertically from the foundation level up to intermediate floors, the roof, or both. Individual wall panel units or portions of a given panel unit can be designed using the "slender wall" provisions as defined in section 14.8 of the American Concrete Institutes (ACI) publication ACI 318 "Building Code Requirements for Structural Concrete". ACI has also developed an expanded guide specifically for tilt-up design through its Committee 551. Publication ACI 551.2 "Design Guide for Tilt-Up Concrete Panels" should be a familiar reference for those involved in the structural design of tilt-up projects or for those who wish to learn more about their design. There is also an industry organization named the Tilt-Up Concrete Association (TCA) that serves as an educational and promotional resource for tilt-up building systems. The TCA now offers a state-of-the-art publication entitled "Engineering Tilt-Up" that is a comprehensive design reference featuring design examples that will undoubtedly become an industry standard as time goes on. The proper design of load bearing tilt-up concrete wall panels must take into account axial compressive loads, primary bending moments due to lateral loads (wind or seismic), primary bending moments due to axial load eccentricities and secondary bending moments due to P-Delta effects. Through a process of trial and error, an assumed combination of panel thickness, concrete strength and proportion of reinforcing steel is investigated. When the resultant stresses and deflections from all of these forces acting simultaneously meet the prescribed code limitations, then the investigated section is deemed adequate.

viii Houston Business Journal, CBRE, and the National Association of Industrial and Office Properties Houston commonly publish annual reviews that this statistic is derived from.

ix Total number of firms affiliated with the Local Houston AIA Chapter 2006.

x   For an example see "Good-Life Modernism and Beyond. The American House in the 1950s and 60s: A Commentary" by Mark Jarzombek (*The Cornell Journal of Architecture*, Vol. 4., Fall 1990).

xi Kelbaugh, Douglas. "Seven Fallacies in Architectural Culture", *Journal of Architectural Education*, Vol. 58, Issue 1, September 2004.

xii Kwinter, Sanford. "La Trahison des Clercs", *Far From Equilibrium: Essays on Technology and Design Culture*, Acrtar Press, Barcelona, 2007.

Tilt Wall as Product and Commodity

An Illustration of Tilt Wall's Potential as Product and Commodity—
The Value Office™ Product Line

This section is both a summary of the argument put forth in the previous chapter and a premonition of the one to be made in the next. In the next, the potential of Tilt Wall to be considered as technology is explored. In contemporary culture, technology is meant to be consumed, turned into product, or commoditized. Tilt Wall may have the very same potential. In the last chapter, a two-part conclusion hinted at Tilt Wall's potential as a research vehicle that would contribute to the arsenal of design opportunities for architects. Summarized, Tilt Wall is an economically accessible way to facilitate the experimental in the general practice, particularly on certain building types sensitive to market conditions. Institutions, in many cases, are more insulated from quarterly market financing and investment pressures, but building types such as office buildings, specifically those built speculatively, can benefit in terms of Tilt Wall's low cost structure, its design and formal flexibility. In normalized markets, on office buildings, Tilt Wall will perform at an average of $10 per square foot less than comparably sized and architecturally appointed conventional constriction. This is due to its lighter overall steel tonnage, as it has no perimeter columns, and faster general conditions on the part of the contractor, thus less time spent under construction. The result has been a rapid adaptation of this "product"—this way of building a comparable office building for less—into a traditionally constructed office market. The following is an illustration of what Tilt Wall office buildings in a new market position can look like.

First, the foundations of the premise of the proposed research have to be outlined. Reduced to almost fundamental simplicity within the discipline of architectural theory, a great deal of debate was held regarding the notion of architecture as a commodity. Particularly in the mid 80s, journals such as the *New Left Review* explored postmodern cultural transformation—particularly in art and the social structure of the public realm—as well as the relationship of representational systems to power, which overlapped with architecture. Much of the debate was structured around reconciling the modern movement's imputed inspiration by Marxist principles with its capitalist aspirations and manifestations. The complexity of this exploration of what was proposed as the commodification tendencies in Modernism and the simultaneous rejection of those same capitalist urges is exemplified below in two quotes:

What has happened is that aesthetic production today has become integrated into commodity production generally: the frantic economic urgency of producing fresh waves of ever more novel seeming goods from clothing to airplanes, at ever greater rates of turnover, now assigns an increasingly essential structural function and position to aesthetic innovation and experimentation ... Architecture is, however of all the arts, that closest constitutively to the economic, with which, in the form of commission and land values, it has a virtually unmediated relationship ...

Fredric Jameson, "Postmodernism, or The Cultural Logic of Late Capitalism", New Left Review, *146, July-August 1984.*

High modernism, as Fredric Jameson has argued elsewhere, was born at a stroke with mass commodity culture. This is a fact about its internal form, not simply about its external history. Modernism is among other things a strategy whereby the work of art resists commodification, holds out against the skin of its teeth those social forces which would degrade it to an exchangeable object. To this extent modernist works are in contradiction with their own material status, self-divided phenomena which deny in their discursive forms their own shabby economic reality. To fend off such reduction to commodity status, the modernist work brackets off the referent or real historical world, thickens its textures and deranges it forms to forestall instant consumability, and draws in its own language protectively around it to become a mysteriously autotelic object, free of all contaminating truck with the real ... But the most devastating irony of all is that in doing this the modernist work escapes from one form of commodification only to fall prey to another. If it avoids the humiliation of becoming an abstract, serialized, instantly exchangeable thing, it does so only by virtue of reproducing that other side of the commodity which is its fetishism.

Terry Eagleton, "Capitalism, Modernism and Postmodernism", New Left Review, *152, July-August 1985.*

CONVENTIONAL CONSTRUCTION

PERIMETER COLUMN

COLUMN FREE

VALUE OFFICE: 2.0 & 4.0

VALUE OFFICE: 3.0, 3.5, 4

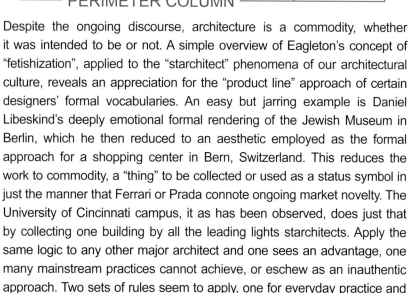

Despite the ongoing discourse, architecture is a commodity, whether it was intended to be or not. A simple overview of Eagleton's concept of "fetishization", applied to the "starchitect" phenomena of our architectural culture, reveals an appreciation for the "product line" approach of certain designers' formal vocabularies. An easy but jarring example is Daniel Libeskind's deeply emotional formal rendering of the Jewish Museum in Berlin, which he then reduced to an aesthetic employed as the formal approach for a shopping center in Bern, Switzerland. This reduces the work to commodity, a "thing" to be collected or used as a status symbol in just the manner that Ferrari or Prada connote ongoing market novelty. The University of Cincinnati campus, it as has been observed, does just that by collecting one building by all the leading lights starchitects. Apply the same logic to any other major architect and one sees an advantage, one many mainstream practices cannot achieve, or eschew as an inauthentic approach. Two sets of rules seem to apply, one for everyday practice and one for avant-garde practice; it is either acceptable to explore or exploit the notion of architecture as commodity or it is not. The quality of that exercise in "architecture as a product" is what really ought to be in question.

Value Office™ is a term trademarked by Powers Brown Architecture. Value Office is the conversion of a typical, conventionally constructed (that is steel or concrete-framed perimeter column, non-load bearing wall construction) market office building to Tilt Wall technology. The original task—one of "commodifying" the building method through a building type—was initially an uphill struggle. We had to prove to commercial

55'

VALUE OFFICE: 5.0

VIEW OF COLUMN FREE CORNER GLAZING

developers that the utilization of a low cost construction technology to build an investment grade office building could compete with the definition of Class A in the market place. The aesthetic flexibility of Tilt Wall combined with its ability to improve upon the characteristics of typical offices ultimately won a niche for Value Office in numerous national market regions. It improved upon the prevailing product (due to no perimeter columns), making it more efficient to plan; in addition, its wall strength accepts any exterior cladding material and it offers column free all-glazed corners—these are only a few of its ultimately appealing attributes. Importantly, when combined with the above advantages, a project build of this kind averages $7 to $10 less per square foot, in most suburban markets.

This represents a grass-roots insurgency of low cost technology, fueled by historic economic forces. It occurs in a space that many architects have occupied—developer work, investment grade commodity architecture—but as contractors and developers have been more open minded regarding adaptations like Tilt Wall into "high end product", it inadvertently threatens to leave that area of architectural practice behind if ignored. As Tilt Wall is a "system", it does not always appear to require an architect. Our effort has been to engage the economic metrics head-on to control the possibility of the aesthetic potential of this building type that makes up so much of the everyday suburban landscape.

VALUE OFFICE 2.0
133,000 SF: Office Building

Overall View of Building Exterior

View of Front Entrance

Lobby

DETAIL OF BUILDING UNDER CONSTRUCTION

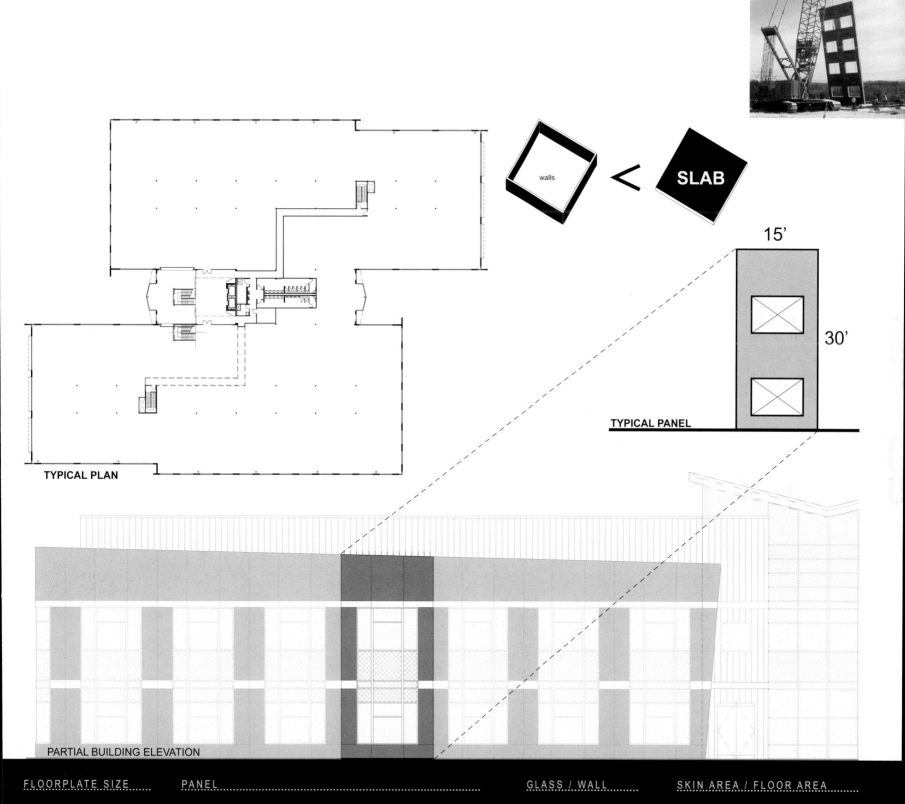

walls < SLAB

15'

30'

TYPICAL PANEL

TYPICAL PLAN

PARTIAL BUILDING ELEVATION

FLOORPLATE SIZE

60,000 SF

PANEL

31,520 lbs
WEIGHT

15 ft X 30 ft
DIMENSION

7 in
THICKNESS

GLASS / WALL

38% / 62%

SKIN AREA / FLOOR AREA

40,800 sf / 120,000 sf

VALUE OFFICE 2.1
150,000 SF: Office Building

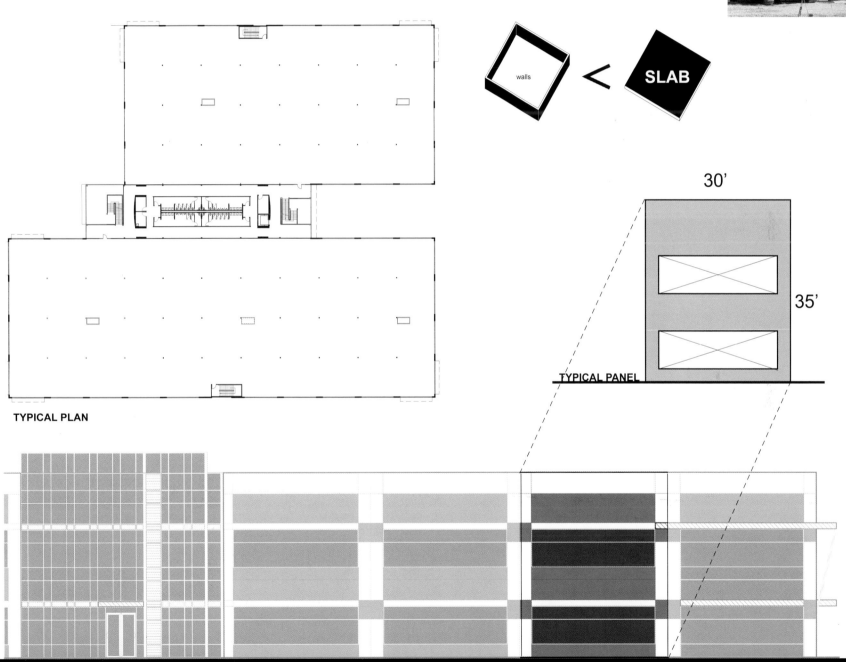

walls < SLAB

30'

35'

TYPICAL PANEL

TYPICAL PLAN

FLOORPLATE SIZE	PANEL			GLASS / WALL	SKIN AREA / FLOOR AREA
75,000_{SF}	**71,240**_{lbs} WEIGHT	**30**_{ft} **x 35**_{ft} DIMENSION	**8**_{in} THICKNESS	**32%** / **68%**	**49,000**_{sf} / **150,000**_{sf}

VALUE OFFICE 3.0
164,000 SF: Office Building

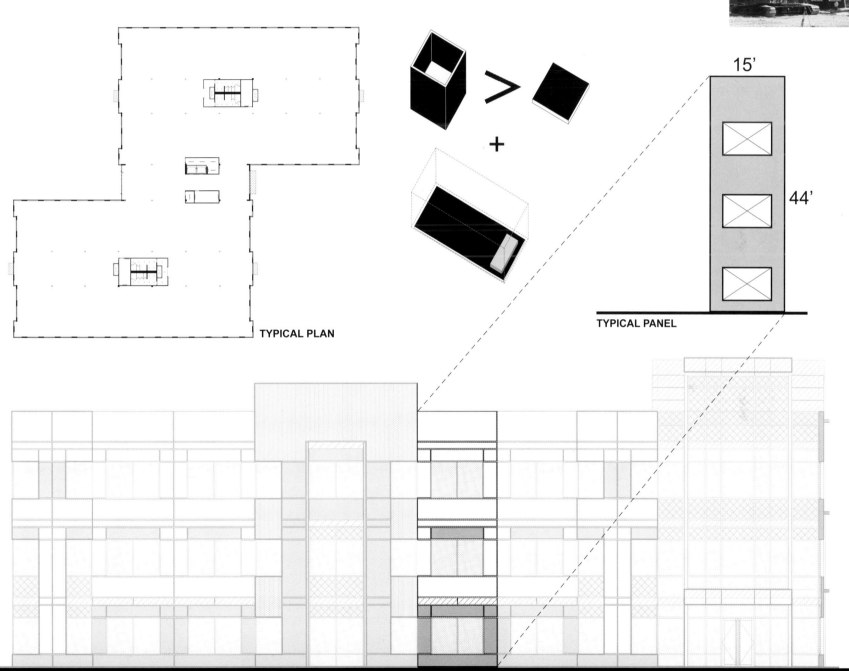

TYPICAL PLAN

TYPICAL PANEL

15'

44'

FLOORPLATE SIZE	PANEL			GLASS / WALL	SKIN AREA / FLOOR AREA
55,000_{SF}	**62,900**_{lbs} WEIGHT	**15**_{ft} **x 44**_{ft} DIMENSION	**9**_{in} THICKNESS	**31**% / **69**%	**51,000**_{sf} / **165,000**_{sf}

VALUE OFFICE 3.1
56,000 SF: Office Building

OVERALL VIEW OF BUILDING UNDER CONSTRUCTION

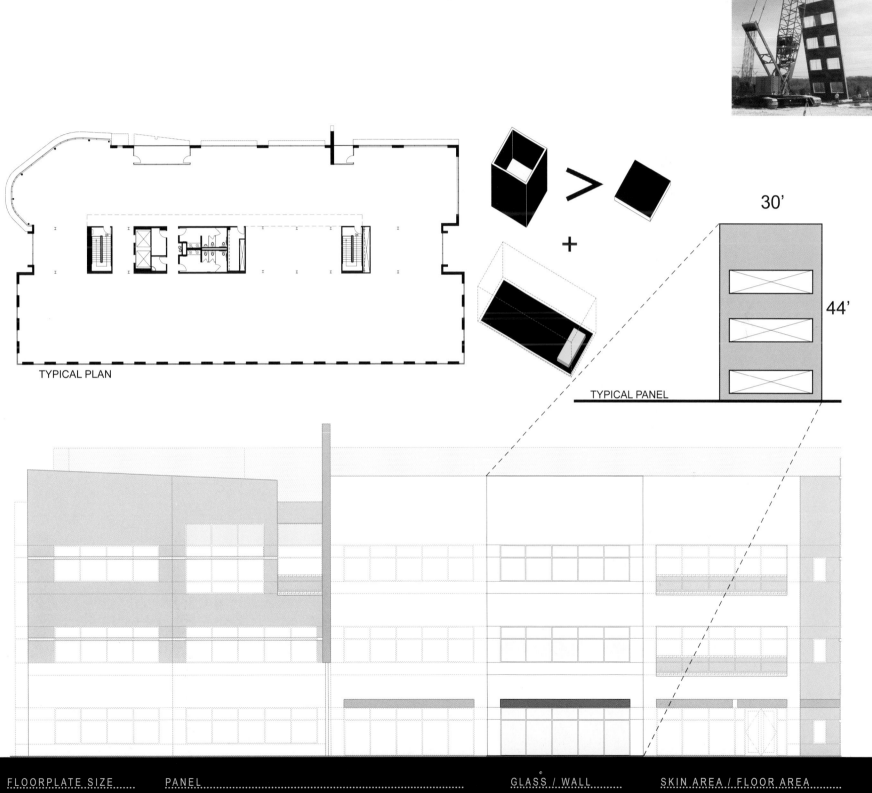

TYPICAL PLAN

30'

44'

TYPICAL PANEL

FLOORPLATE SIZE

15,000SF

PANEL

101,680lbs
WEIGHT

30ft x **44**ft
DIMENSION

9in
THICKNESS

GLASS / WALL

38% / **62**%

SKIN AREA / FLOOR AREA

27,720sf / **45,000**sf

VALUE OFFICE 4.0
135,000 SF: Office Building

PHASE **02** TO MATCH

PHASE **01** BY OTHERS
(CONVENTIONAL CONSTRUCTION)

VIEW OF BUILDING UNDER CONSTRUCTION

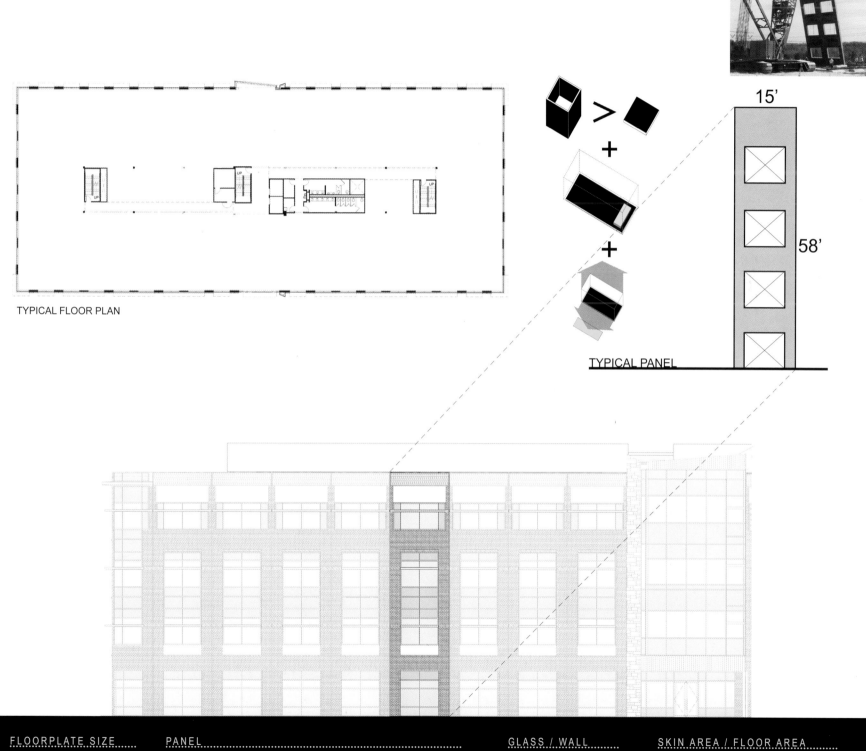

TYPICAL FLOOR PLAN

15'

58'

TYPICAL PANEL

FLOORPLATE SIZE

32,000SF

PANEL

72,160lbs
WEIGHT

15ft **x 58**ft
DIMENSION

12in
THICKNESS

GLASS / WALL

34% / **66**%

SKIN AREA / FLOOR AREA

47,000sf / **128,000**sf

Post Great Recession, no business, profession or trade will take up where it left off before this historic economic event. There is a new normal. Architects, however, give every indication that they simply need to wait until the market "comes back". When it does, it will not come back to them.

They seem to misunderstand the new normal. Architects will have to "go to" the new market with a product rather than wait for the market to request their services. The value proposition of architecture has changed. There are no more clients, only markets now. Innovation in mainstream everyday practice requires that we understand markets don't behave like clients … they are open 24 / 7 / 365 … and unlike clients, markets can be created where none existed before.

To build a market you have to have a product.

Or the notion that you can commoditize something—even if it appears to be ubiquitous, it may NOT be commodity … yet …

Architects, engineers and contractors don't really like hearing the buildings they pour themselves into referred to as Commodities,

Get over it.

All it really means is realizing the commercial potential of what is currently being perceived as a REQUIREMENT.

You don't have to compromise on integrity, purpose or the poetry of your soul.

Tilt Wall construction's capacity to instigate and sustain the investigation of transformative modalities marks its difference in potential from competing value-oriented building technologies. Balancing high design and technical innovation with form driven construction is unique to Tilt Wall.

The "new" is often only a subtle transformation of the old, disguised as the radical … re-packaged … often based on similar fundamentals …

Sometimes what appears to be radically new is really, in essence, not that radical. It is merely a repackaging to suit the times—we put our effort into imagining what can constitute "new" in an office building—certainly one of the most ubiquitous and formulated of all building types. In the context of Tilt Wall—apparently quite a bit.

VALUE OFFICE 3.5

170,000 SF: Administrative Facility

> (diagram) + (diagram)

15'

58'

15'

44'

TYPICAL PANEL

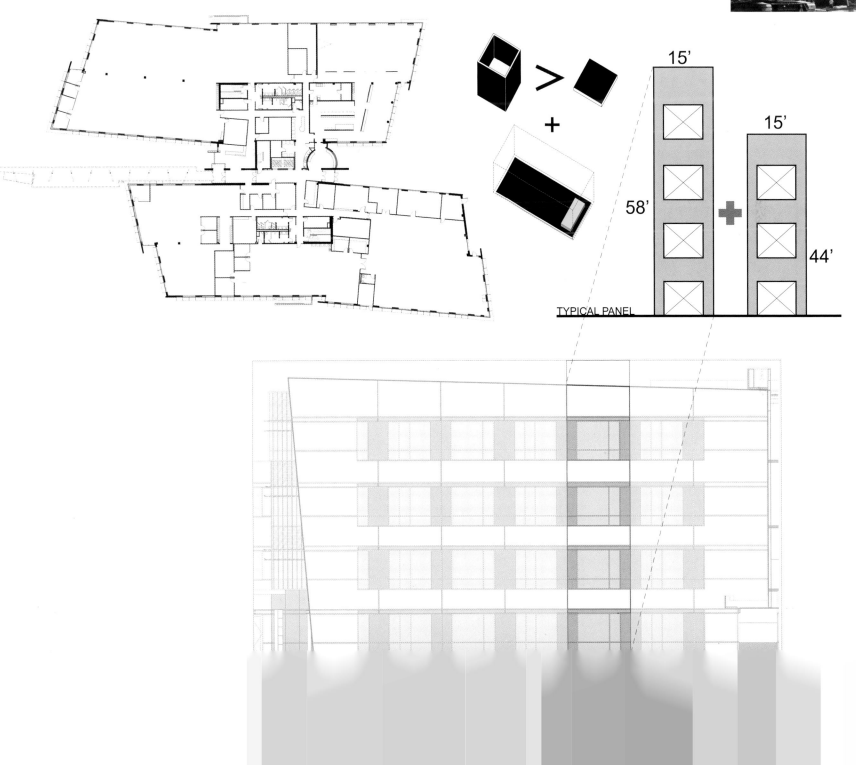

VALUE OFFICE 4.1
240,000 SF: Office Building

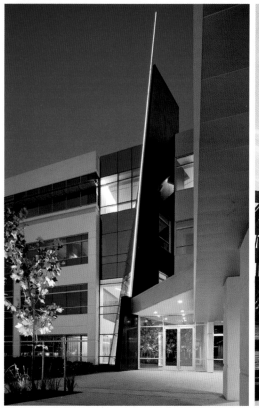

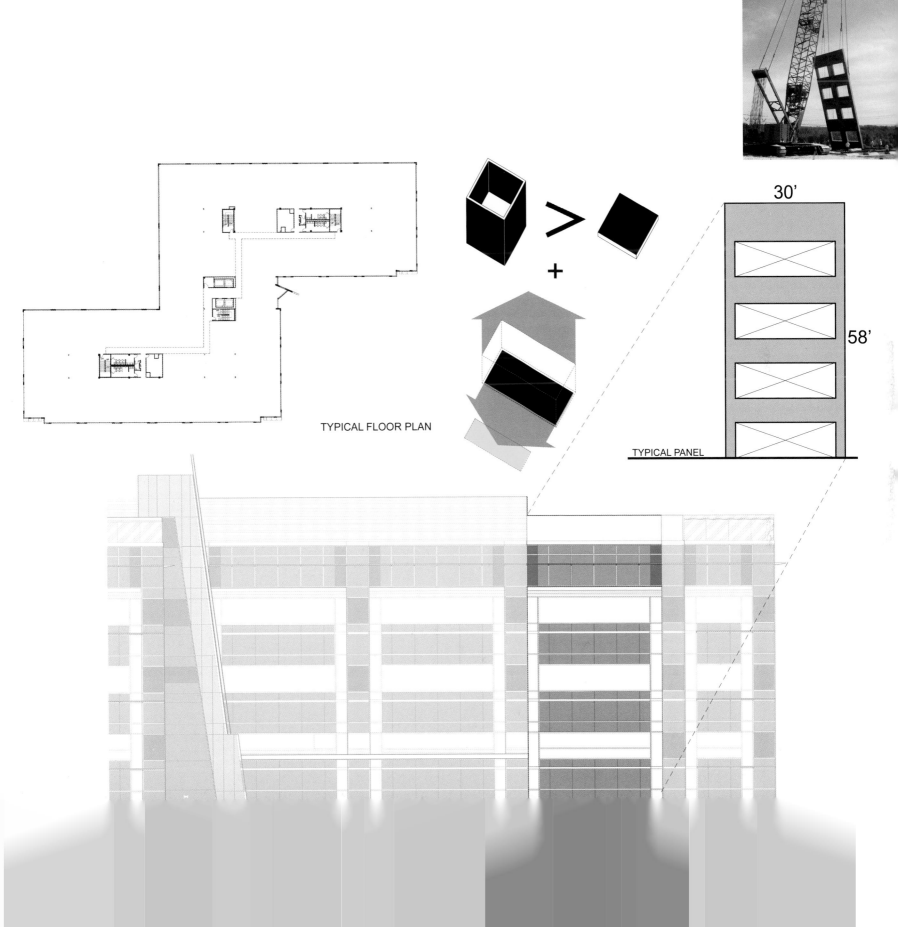

TYPICAL FLOOR PLAN

30'

58'

TYPICAL PANEL

VALUE OFFICE 5.0
240,000 SF: Office Building

TYPICAL FLOOR PLAN

60'

16'

56'

TYPICAL PANEL(S)

30'

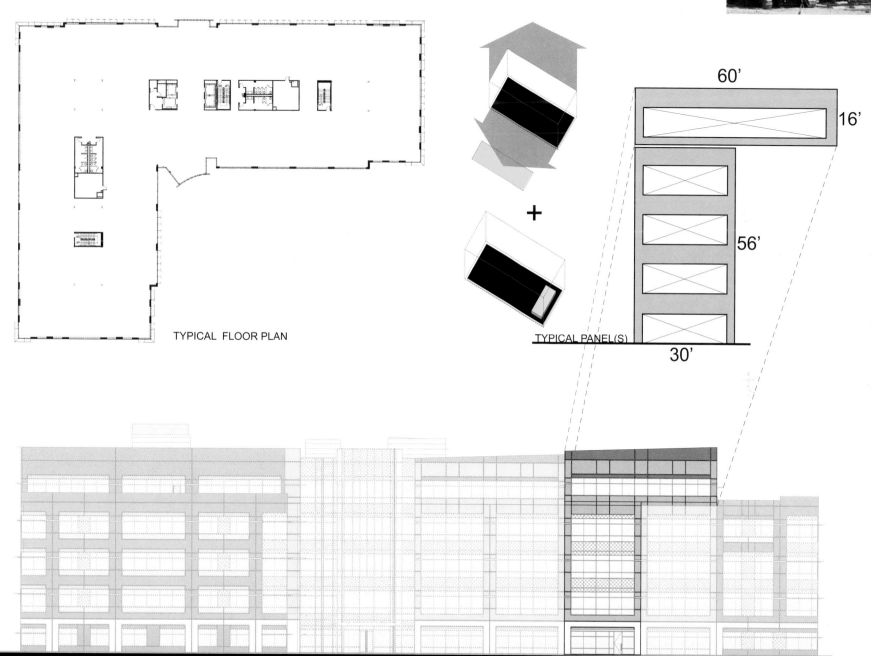

FLOORPLATE SIZE **50,000**SF

PANEL

148,000lbs **60**ft x **16**ft **7**in TOP +
+ 51,500lbs **+ 30**ft x **56**ft **12**in BOTTOM
WEIGHT DIMENSION THICKNESS

GLASS / WALL .... **41**% / **59**%

SKIN AREA / FLOOR AREA .... **80,160**sf / **241,200**sf

VALUE OFFICE 6.0
162,000 SF: Office Building

OVERALL VIEW OF BUILDING

Typical Panels

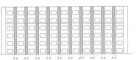

5'–0" Spread

4–2 / 10'–0" Shift

**20'–0" Panels w/ 2'–6"
Cantilever on each Side**

Center-Loaded Entry

Primary
Typical Window-Opening Pattern

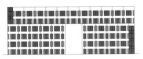

Secondary
2'–6" vs. 5'–0" Glass

Tertiary
Horizontal Sunshades

Tertiary
Vertical Sunshades

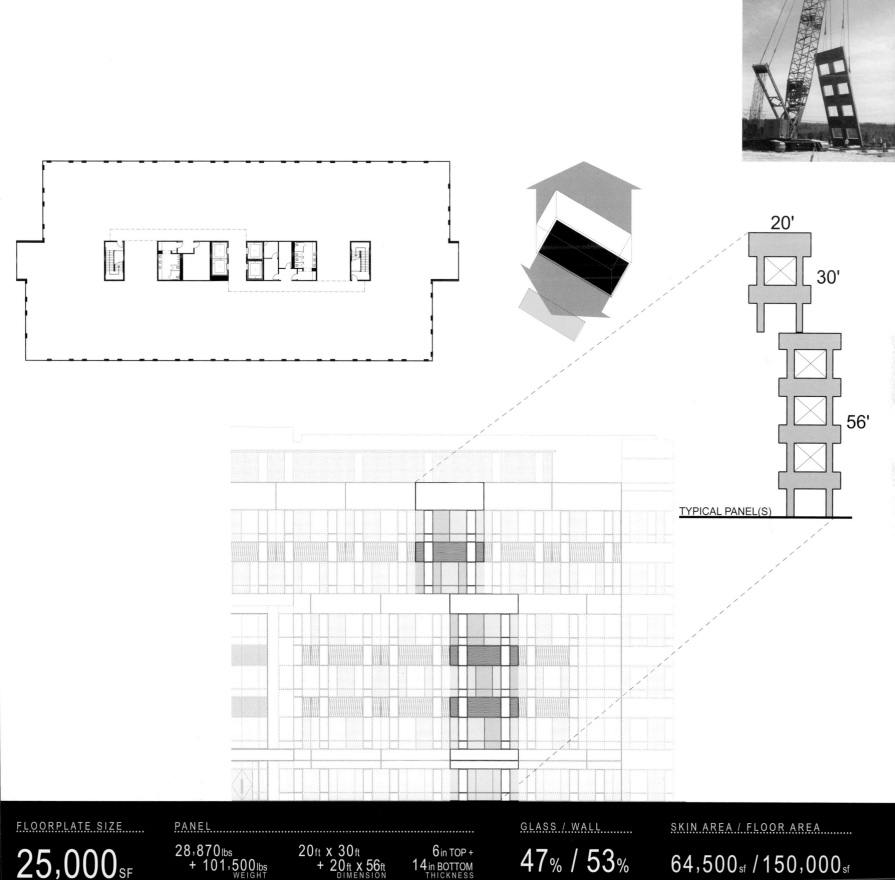

20'

30'

56'

TYPICAL PANEL(S)

FLOORPLATE SIZE .....

25,000SF

PANEL

28,870lbs
+ 101,500lbs
WEIGHT

20ft x 30ft
+ 20ft x 56ft
DIMENSION

6in TOP +
14in BOTTOM
THICKNESS

GLASS / WALL ....

47% / 53%

SKIN AREA / FLOOR AREA .....

64,500sf **/ 150,000**sf

Tilt Wall as a Potential Carrier of Meaning
05

"For Heidegger, technology is problematic not in regards to benefit it brings but rather to its emergence as an autonomous force ... "

Kenneth Frampton, Studies in Tectonic Culture: The Poetics of Construction in Nineteenth and Twentieth Century Architecture.

"We continue to hear the fiction that what is important is how a technology is used ... but hardly ever do we mediate on wider implications: on what a given technology, or group of technologies ... actually does to us"

Sanford Kwinter, Far from Equilibrium: Essays on Technology and Design Culture.

Carrier of meaning = theory = research

The goal of this section is to establish the relationship between architecture as an embodiment of, or carrier of meaning, and how it does so, namely relating to the concept of tectonics. In particular, this section is an exploration of the latent content in the Tilt Wall construction methodology. How form is made either contributes to a larger argument for content and meaning in a formal proposition, or it doesn't. There appears to be no middle ground. The relationship between process and content in Tilt Wall construction must reside in its being a tectonic method and thus a potential contributor to subjective intention, rather than its categorization as technology, and thus a neutral objective technique. The connection between construction methodology and formal meaning necessarily travels through the discipline of architectural theory. As to why Tilt Wall would present a special case under this rubric, *Learning from Las Vegas* serves as a good model. Venturi, writing as an architect and not a theorist, exposed a previously unseen urban prototype lying on the desert floor. He, along with Denise Scott Brown, Steven Izenour and their Yale students, teased the latent content out of the (at that point) non-architecture of the Vegas strip: They threw into sharp critical relief buildings by unimportant architect's containing previously non-existent programs, which were running as an independent, bypassing system of meaning relative to conventional architectural culture. Tilt Wall construction is comparable.

Tilt Wall is simply a method of building and thus arguably empty of content. Yet we have always attempted to assign meaning to how we build, from the formulation of the primitive hut trope, to the cataloging of form in the Enlightenment Encyclopedias, to Gottfried Semper's primitive hut modifications involving cladding principles. More philosophically José Ortega y Gasset's definition of constructing a cannon, which connects making, form and meaning: "For in truth, the most accurate definition of the urbs and the polis is very like the comic definition of a cannon. You take a hole, wrap some steel wire tightly round it, and that's your cannon". Modernist architecture's protagonists, both practicing and proselytizing, were partial to mass production as a way of building or conceiving of form, a notion that has recently been recycled in its latest iteration as remanufacturing, where the process of production bears meaning in the production of form.

The relationship to how we build and what the final form means has been negotiated by formalized theory in its many guises. Theory proper came into its own from the late 70s to the early 90s by channeling numerous concepts including language structure, French Structuralism and Marxist arguments on capitalism and art. Evaluated was: the discipline of writing within architecture, speculation about the relationship of architecture to cultural superstructures, how it carries or generates meaning, and why it does so. That architecture can be debated in terms of, or even as, a theory / theoretical proposition has been a questionable construct for many architects. Architects are uncomfortable with this concept and, in fact, the theory game has been declared by many to have run its course in the last five years or so.[i] On the other hand, architecture continues to rely upon decoding or interpretation; it seems to need to be understood as a work of subjective authorship. As such, it has always been a purveyor or indicator of cultural intent. Therefore, discussions about what it comes to mean, and examining whether architecture is a kind of "midwife" to cultural content, or an interpretable analog or index to a process of its own becoming, are a useful and required discourse whether or not such discussions congress into theory. Diane Ghirardo's work on the social potential and responsibility of architecture in the early 90s is one such discussion that, in many ways, applies to Tilt Wall construction's architectural value. In her book *Out of Site: A Social Criticism of Architecture* Ghirardo says, "the architect can engage critically with contemporary problems solely through formal manipulation", in other words, that through form alone, one could contest things such as the commodification and consumption of culture. She argues that a new notion of architecture, oscillating between self-expression and effete cultural commentary, replaced the traditional "architect as critical interventionist" program of social, economic and political content. To facilitate this, architects and critics conspired to create a rubric for evaluating buildings, which in turn became the architect's conceptual framework—all discourse not in this closed system is excluded and critics only address, in the end, areas in which architects have had some input. Thus questions regarding economic consequences, social implications and so on, can (and indeed have) been carefully edited out of critical consideration. In short Ghirardo concludes that entire discourses over questions such as building processes and methods are written out of "critical" architecture in favor of correct surface

MODERNISM

POSTMODERNISM

STRUCTURALISM

POSTSTRUCTURALISM

BLOBISM

BIOMORPHISM

CONSTRUCTIVISM

DECONSTRUCTIVISM

MINIMALISM

BRUTALISM

CLASSICISM

ECLECTICISM

FUNCTIONALISM

FORMALISM

EXPRESSIONISM

REALISM

URBANISM

FUTURISM

NATURALISM

ROMANTICISM

TILTWALLISM

manipulation and the latest "ism". In this context it is easy to see that the exploration of market-based methodologies of form making, like Tilt Wall construction, have been systematically downgraded as integral contributors to formal content and meaning.

Ignored by the new "research"

Theory has its problems as a catch-all discipline for understanding arguments outside of its own constructs. But this autonomous self-referential content was not always what constituted theory. Well into the 60s architectural theory was simply architectural history. The transformation of history into a conduit for outside discourses was not seamless, and had a seismic effect, yet theory in this state has proven to be somewhat short-lived. Another shift has recently occurred in which theory has been transformed into *research* in the majority of academies, and in curricula. In its many manifestations, from the instrumental digital-fabrication labs to the data-driven performance conditions of sustainability, research has, in turn, displaced building as the vehicle of architectural manifestation. So, what is left of structured, institutionalized academic theory? It now acts as the basis of the enthusiasm of architectural academics for research. In so transforming, it has gone from the broad and cultural scope to the narrow and instrumental set piece. This development appears, at first, as an opportunity for tectonic methods like Tilt Wall as relates to a current regulatory discourse of meaning in architecture. Instrumental research into: the chemical composition of concrete; the effects of alternate concrete formulations and products, including aerated autoclaving; low-flow concrete, and flowable plastics and polymers, are circumscribed by Tilt Wall as a heuristic technique. But research has thus far concentrated on the complicated, the small, and the non-synthetic topics, rather than something as rigid as a "technique". There are numerous reasons why a narrowing of subjects should be emphasized over more ecumenical approaches to research.

Architecture as an academic discipline seemed, for the longest time, to resist fundamental academic research. In a reversal of the accepted format of exchange between academia and practice (where one experiments in *school*, and *practice*, slothfully and with great resistance to impracticality,

is influenced by the results), practice took up the mantle of "research" and has ironically pushed the concept back into academia.[ii] Ten years ago the word research was not to be uttered in architectural schools beyond historical PhD candidates. It was implicit that buildings in themselves were the ultimate form of architectural research. But no longer. The recent origins of practice-based research are not obscure; they are informed by Venturi's work in *Learning from Las Vegas*, although a caesura seems to have elapsed. OMA and the Dutch school of data-driven form are successful "re-purposers" of the 60s "waiting for output"[iii] pseudo-research architects. The almost instant academic legitimacy of research is illustrated by the most recent advertisement from a prominent academic organization's (ARCC) conference brief:

> Recent decades have witnessed a notable expansion of architectural research activities, with respect to both subject and methodology. This expansion can be mostly credited to an increase in government and private funding of primarily academic research initiatives. More recently, however, a noticeable increase in research activities within the architectural profession makes it possible to argue that it is the profession itself that is now taking leadership in the development of contemporary research agendas …

In some ways it seems preposterous that, in such a short time, architecture has become the discipline guiding the forefront of more general research agendas. But it is clear that theory has given way to research as the hegemonic academic activity. Regarding research—as it has become defined and appropriated by architecture—Tilt Wall fails on two fronts: it is not high-tech, and, as it largely relates to the construction of buildings, it is too large scale. That is, the architecture of *buildings* seems to have been trapped in the no-man's land of scale as circumscribed by current academic activity. Research, as most commonly executed, is small scale, almost installation scale; akin to David Niland's definition of architecture as "art with toilets".[iv] Much of what is described as research does not even bother with toilets. Digital fabrication, "performative" building components (especially facades), the invention of new materials (translucent concrete—but what is to become of glass?), the integration

Architect Magazine: **Entry for R+D Awards**

New technologies are revolutionizing the process and product of architecture. To celebrate advances in building technology, *ARCHITECT* magazine announces its seventh annual R+D Awards. The awards honor innovative materials and systems at every scale—from entire buildings to HVAC and structural systems, discrete building materials such as wood composites and textiles, and digital tools such as design software and mobile apps. The R+D Awards welcome building technologies of all types and encourage the broadest possible dialogue among architects, engineers, manufacturers, researchers, students, and designers of all disciplines.

Categories
→ Prototype
  Products, materials, systems, and software that are in the prototyping and testing phase

→ Production
  Products, materials, systems, and software that are available for use

→ Application
  Products, materials, systems, and software as used in a single architectural project or group of related architectural projects

The jury will consider new products, materials, systems, and software as well as cutting-edge uses of existing products, materials, systems, and software. Software entries may include digital technologies such as programs, mobile apps, and cloud-based tools. Entries will be judged for their potential or documented innovation in fabrication, assembly, installation, application, execution, usability, and performance. The jury will also judge entries on their potential to advance the aesthetic, environmental, social, and technological value of architecture.

of new technologies (LED), all to be processed by B.I.M. and cut with lasers, CNC routers and 3D printers; all of the above most often describe the very small scale in architecture. Certainly there is rumor of research at the very large scale as well. One is led to conclude, from the copious amount of critical literature, that urban design is not yet dead. Professional trade publications featuring Michael Speaks and Michael Sorkin and the recent Michigan Debates on Urbanism led by Douglas Kelbaugh all take second seat to Rem Koolhaas's Project on the City as acts of research. Still, architecture is most often attempted at the scale of buildings, and research is not. Tilt Wall construction occupies the former category but lacks, at least without investigation into material elaboration and variation, the apparently progressive cultural spirit that lasers seem to capture. Our collective fascination with the new, our facile maneuvering between the zeitgeist and the avant-garde, endangers our claim on what has been understood to be exclusively our territory—*the building*. We seem quite happy to submit ourselves to building only parts, or to strategizing the effects of how they aggregate to form communities and/or retard minority politics.[v] The mechanisms resulting from research activities are then employed by others: "mainstream" architectural practices polluted by market willingness, or worse, builders aping the conclusions in low-tech Tilt Wall (think of New Urbanism or any "Life Style" center). Architecture's insistence on exploring new territory as defined by beguiling technology is what relegates Tilt Wall construction to a kind of neglected territory. We are, for reasons unknown, bound to the traditions of the avant-garde and to the myth of the zeitgeist. It seems in our "research" we are compelled by them and, taken to an extreme, without critical interference, one would come to the conclusion that the blog is the highest form of literary achievement we as a culture have produced. Many—none more coherently than Preston Scott Cohen in his recent essay "The Hidden Core of Architecture"— have begun to attempt to reconcile this dilemma of research's limited scale and the need to incorporate its results into building, in ways inclusive of Tilt Wall as a system, or at very least inviting its study.[vi]

Scott Cohen observes two kinds of architects: those who manipulate the surface and those who manipulate the spatial properties of the core. Thus he has divided architecture into what he defines as the more permanent elements of core, such as structure and vertical circulation and

so on, from, simply put, cladding. In so doing he describes a dismantled tectonic unity, where the skin invites one set of critical operations (fast / temporary) and the core another (slow / permanent). He goes on to conclude that computation (read here research) has simply created new styles of cladding the surfaces of cores. In fact, the skin of buildings, he posits, have become the recipients of the benefits of research; they are fast, temporary and transient, often conceived of as "installations" to be dismantled in not too distant futures and replaced by more up-to-date efforts. Research is, as argued and confirmed by Scott Cohen, partial and intransient, impermanent and therefore not really transformative: "the thematization of temporariness facilitates legitimization of the ziegeist".[vii] He appears to confirm the problem of research in and of itself while also inviting synthetic tectonic systems such as Tilt Wall as possible solutions. After all, what is Tilt Wall if not the convergence of infrastructure (core) and skin? It occupies simultaneous roles, supporting permanent structure and concomitant communicating surface. It resists the disposability of so many of the avant-garde propositions Scott Cohen laments, while engendering the very quality which observations such as his invite. Certainly the application of Tilt Wall construction to his provocation is convenient and not here proposed as an off-the-shelf "solution". Rather the present instigation is to illustrate just how convenient current critical thinking and the study of tectonics like Tilt Wall appear to be in the ongoing critical discourse.

Can research and the academy afford to exclude areas that do not comport with the high-tech hubris that defines progress? Is Tilt Wall just too crude to be subject to intense scrutiny and potential evolution? That is to say, is it an end in itself? It seems that in order to legitimize certain forms of research, the building must maintain its traditional importance as the object of architectural manipulation. Buildings, of course, can be the recipients of the positive outcomes of research as described by the small (technology) and the large (culture), but they also have the potential to add so much more to the limited benefits of current research. Tilt Wall construction deals exclusively with the act of building and must be included as a valid venue of investigation. Proposing just how Tilt Wall construction can benefit designers is being defined more precisely as a subject for architectural research by situating it within a broader context of its status as a constructive medium.

Technology vs. tectonics

The beginning of this section granted Tilt Wall construction inclusion under the nomenclature of tectonics. In so doing it skipped over discussions involving technology, a broader category of cultural concern that overlaps architectural discourse more specifically than tectonics. So why is Tilt Wall a tectonic system rather than a technology? Or is it both?

As a pretext to legitimizing Tilt Wall as a meaningful tectonic method, it is first necessary to situate it within the realm of technology and, in particular, architectural technology, then to understand the current relationship of architectural meaning as a byproduct of technology. Antoine Picon, in an essay entitled "Does Our Technology Make the Past Irrelevant to Our Future?" in *Harvard Design Magazine*,[viii] defines technology:

> Technology is a term so widely used that it has migrated to an all circumscribing meaninglessness very much like the words "Modernism" or "Program" and thus requires a specific definition of intended use in this context. Roughly stated, technology here refers to machines, devices and contrivances that increase the apparent quality of existence.

Picon's essay clearly outlined the history of technology's decline from dependence upon the past as an authenticator. That is to say he has described the de-coupling of technology from progress, and its manifestation as a reflection of society and a progenitor of meaning / progress. In short it has become a form of hubris realigning architecture and technology in a meaningful way; indeed Picon posits a qualified and redefined relationship with the past as a non-progressive way to move forward: "Why not consider instead that technological development may have to do with contingency, even randomness?" Jose Ortega y Gasset similarly argues that redundant technologies that serve the same ends undercut the notion of progress.

This is precisely the context in which Tilt Wall, as a potentially ripe source of architectural exploration, is allowed through the door. For what is it if not a vestigial technology and, simultaneously, a currently valid one? Vestigial at least in so far as it remains currently valid as a widely popular commercial vernacular approach to building. This current popularity persists, despite

Tilt Wall having been "invented" as a progressive building technique over one hundred years ago; a condition similar to many of the techniques of general building methodologies—very little "progress" has been made in the fundamentals of how we build. The equipment we build with has changed technologically but the process itself remains static. Does Tilt Wall not possess something of the validity of the past—as a form of the Industrial Revolution's progress—and simultaneously a freedom from it? Could it not be capitalized upon given technology's current focus on digital form making and sustainable measurement, versus the literal making of and construction of form?

So why not accept that Tilt Wall is a technology? By analogy, one accepts that a cell phone is technology, but rarely considers the quantum physics of the microchip that makes it work as technology. It has been argued that technology should not be used to measure the correctness of form without parameters, within which it must be understood—thus, it's reduction to technique. So, the microchip is the technique on which the technology of the cell phone works, or has progressed. If Tilt Wall is defined as a mere technique, it is limited to being understood as only an aspect of building technology: a distant cousin to "podgiting"[ix] or running bond. We must define the architectural difference between technique and technology as tectonics. The matter is a simple one: technology relates in most sources to the "system"[x] of an activity, while technique is used to describe the execution of a "part"[xi] of a system. If we accept that there are styles of design and there are styles of building in architecture,[xii] tectonics becomes the logical destination for the positioning of Tilt Wall as a viable modus for design exploration. Tectonics combines the act with the intent. It is in effect a style of building. Frampton's *Studies in Tectonic Culture* captures the sense in which Tilt Wall might be considered tectonically:

> Without wishing to deny the volumetric character of architectural form, this study seeks to mediate and enrich the priority given to space by a reconsideration of the constructional and structural modes which, of necessity, it has to be achieved. Needless to say, I am not alluding to the mere revelation of constructional technique but rather to its expressive potential.

Given Frampton's interchangeable use of technique and technology, the issue of Tilt Wall construction's potential expressivity comports with the notion of tectonics as the combination of structure *and* construction as a vehicle of intention and meaning. Tilt Wall, as discussed above in reference to Scott Cohen's skin / core disunity, is simultaneously structure and construction and thus inherently tectonic as a standalone proposition. It is not reducible to merely a "system of construction"; it is an integrated part of a system. It is a part because it alone does not meet the definition of a system of activity (technology) as well as the combination of structural technique unavoidably and inherently combined with expressive potential. Architecture exists in the entirety of its productive processes. Thus by way of concluding analogy, in architecture one must consider the drawing / the computer rendering or indeed the digital model as manifestations of or by products of technology. They lack material reification no matter what level of verisimilitude they obtain to. Technology alone implies a lack of intention and a level of abstraction that is analogous to formal conception on software. It de-couples the physical from the intellectual. Tectonics implies just the opposite. It is the synthetic co-occurrence of intent and physicality. In a brick wall, the mortar thickness and shape confer the technological; the choice of running bond implies the tectonic. Tilt Wall construction has its tectonic legitimacy posited in the inability to separate, as stated by Rowe earlier, the physique and the morale.

Ignored by critical practice

Just as theory's dissolution into research has been a retardant to Tilt Wall's integration into the academy, other forces have conspired to exclude it from serious consideration in practice. The relatively meager interest exhibited by leading-edge architecture practitioners is caused by the historical development of Tilt Wall construction as the by-product of engineering efficiency, not as the medium of any desirable cultural repository.[xiii] There are current opportunities for economic pressures to sponsor architectural meaning—both mundane and controversial. Meaning that is defined by anything more than its alliance with capitalist infrastructure has to exceed Sanford Kwinter's admonition on some Dutch architects' apparent surrender to market dominance.[xiv] Certainly other methods of building and engineering obtain to efficiency and economy, but none as aesthetically

unexploited as Tilt Wall construction. Tilt Wall construction does not, at first glance, appear to be a reasonable method for constructing a museum.[xv] Yet, is it the complexity of the spatial experience a building provides, or the appearance of complexity in the construction we value more?[xvi] Because Tilt Wall construction is boldly driven by its economy, at first glance it is a commodity with almost no pretension to any kind of "meaning".

This, of course, is both its attraction globally (it is "cheap" and easy to perform in low-tech labor pools) and its challenge. It is not an "indigenous", or for that matter local, traditional, or any other culturally pre-loaded approach to building. It evolved as a method of building almost completely outside of accepted cultural expression and in the shadows of traditional architecturally relevant technologies.[xvii] It was driven by the engineering impulses of expediency and simplification. Tilt Wall was, in effect, conceived by Aiken as a commodity, or at least his first act was to make it a product. Frampton's notion of Critical Regionalism,[xviii] simplified to circumscribe the regionalization of modernism by local cultures on a global basis, really does not account for Tilt Wall's development. It does not, in the context of his construct, express any culturally-significant DNA. At the time of Frampton's original argument, there was a greater resistance to commodity and commoditization. This product-like nature of commodification is explored by David Brooks in his essay "Chasing the Storm".[xix] Brooks argues (similar to Frampton on Modernism) that the internet does not operate as an agent of homogenization; it has actually become a source of the mass personal customization of news and information. In effect it is becoming a force for fragmentation, rather than one of consolidation. More recent theoretical developments allow for the validity of "commodity" as a conveyor of architectural meaning, authenticated by the transfer to "projectivity" as the catalyst for meaning or content over "criticality".[xx] As yet, however, this has done little to attract interest in Tilt Wall construction as an avenue of meaningful approach to critical architecture by leading practices.

Tilt Wall is an untapped source of architectural potential. It needs to be positioned as something other than the route to expedient construction, and has to be seen as a contributor to meaning both as a technology and a culturally significant process. A building that is "clear" or honest in terms of its structure has always been an attractive myth, but never the guarantor of anything beyond mere construction. If Tilt Wall is to contribute meaning and thus become a source of more intense research, what it offers has to have a value of acceptable currency: cheapness is not enough. Beyond economy, it has great flexibility in how it "represents", and in how it is positioned and manipulated as a matter of a larger tectonic argument in the scheme of a given solution.

In the vast market created by the prolific use of Tilt Wall construction, much of it in the realm of building outside of architecture, is there, like the casinos on the desert floor, a latent potential that is useful and of interest to architects? If exposed to the scrutiny and investigation of research and experimentation, might not a meaningful expressive system evolve from it? What is needed is a more tectonic consideration of Tilt Wall's potential contribution to what Colin Rowe coined as the "physique and morale" required for meaningful architectural manifestations. That is where the rich territory for exploration lies.

i   Cramer, Ned. "The Silly and the Profound", *ARCHITECT*. October 2013.

ii   For an example see "Once Theory to Practice, Now Practice to Theory?" by Scott Johnson (*Harvard Design Magazine*, Vol. 33, Fall / Winter 2010-2011).

iii   Woods, Shadrach. "Waiting for Printout", *Perspecta: The Yale Architectural Journal*, No. 12, 1969.

iv   Attributed to David Niland. The author attended one of his lectures in 1988.

v   Sorkin, Michael. "The Ends of Urban Design", *Harvard Design Magazine*, Vol. 26, Fall / Winter 2007.

vi   Scott Cohen, Preston. "The Hidden Core of Architecture", *Harvard Design Magazine*, Vol. 35, Winter / Spring 2012.

vii   Ibid.

viii   *Harvard Design Magazine*, Vol. 31, Fall / Winter 2009-2010.

ix   The author observed this common plaster finishing technique in England.

x   *Oxford English Dictionary*.

xi   Ibid.

xii   Ford, Edward R. "Introduction", *The Details of Modern Architecture*, Cambridge: MIT Press, 1990.

xiii   Recently there has been (due to the sheer ubiquity of the method) some limited engagement of Tilt Wall construction by notable architects. Steven Holl utilized it in his Chapel of St. Ignatius, Scogin, Elam and Bray employed it on a public library in Atlanta, Elliott + Associates Architects used it to create a Tilt Wall contractors office in Oklahoma, cunningham architects (sic) have completed a few buildings utilizing Tilt Wall in the Dallas Metroplex (one of which made the national press) and Carlos Jiménez in Houston has experimented with it in a nationally-published project for Cummins Inc.

xiv   Kwinter, Sanford. "La Trahison des Clercs", *Far From Equilibrium: Essays on Technology and Design Culture*, Actar Press, Barcelona, 2007.

xv   That does not mean it has not; see Chipperfield's extension at the Saint Louis Art museum.

xvi   Sam Mockbee and the Rural Studio explored this relationship.

xvii   See Chapter 1.

xviii   Frampton, Kenneth. *Modern Architecture: A Critical History*, Thames and Hudson, London, 1985.

xix   Brooks, David. "Cellphones, Texts and Lovers", *The New York Times*, November, 2009.

xx   Baird, George. "'Criticality' and Its Discontents", *Harvard Design Magazine*, Vol. 21, Fall / Winter 2004.

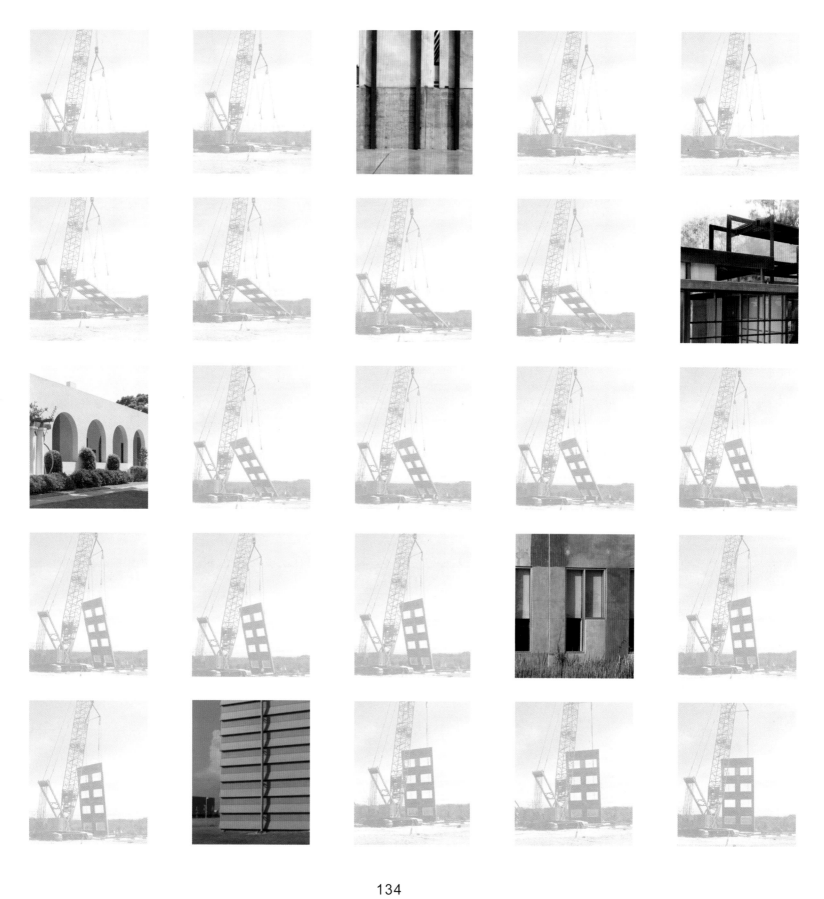

EXCURSUS C
Case Studies

The premise of the argument thus far has been to provoke further interest in the potential of building with Tilt Wall by exploring its implications within larger fields such as design potential, meaning and technology. Much of the concern is that for a proliferating method of construction, it has received a dearth of mainstream and vanguard attention. But that is not to say it has received none. The following Tilt Wall projects have all made the national media or been covered with wide bandwidth interest. The efficacy and purposeful integration of Tilt Wall differs from designer to designer, but this represent the potential of the method in the hands of talented architects.

Irving Gill—La Jolla Women's Club

Irving Gill was unique in his interest in overcoming the exclusion of architects from everyday building types, and the symmetry with contemporary trends is stunning. With the purpose of building more economically and more quickly in order to compete with the then dominant contractors, Gill purchased the rights in 1912 to the by-then bankrupt Aiken Reinforced Concrete Company. Gill first used Tilt Wall on the Banning House in 1913 to the amazement of neighbors, according to *Sunset Magazine* at the time. In 1914, Gill formed a business partnership with his nephew, Louis Gill, and renamed the company the Concrete Building and Investment Company. Its focus was to build low-cost housing. There is, of course, a parallel in that Aiken also utilized the method for low-cost (in his case barracks) housing and that both men saw in Tilt Wall construction the intersection of architecture and business. In some ways, they were early John Portman-types, with Gill exhibiting a modern developer's sensibility. While Gill has been accepted into architectural history more for his early adaptation of Modernism in regionally inflected work, significant in his opus is the La Jolla Women's Club and Clubhouse, for which he utilized the Tilt Wall method to conceive of and construct just prior to forming the building company.

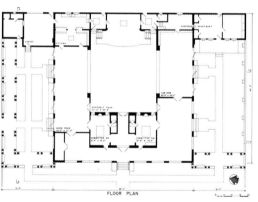

FLOOR PLAN

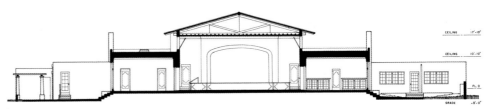

Rudolph Schindler—Kings Road House

While Gill had the aesthetic implications of Modernism in mind with construction playing a secondary role in his work, Schindler's work is defined and characterized by his search for the moral depths of Modernism; the relationship to form and craft were important areas of his investigation. Lloyd Wright (with whom Schindler had worked during his time at Taliesin, in Frank Lloyd Wright's employ) introduced Schindler to Gill and his work. Consequently, Schindler recorded some of Gill's Mission-Style work while on a tour in California in the early 20s, and visited Gill in his office. It is there that Schindler presumably became aware of the Tilt Wall method. He employed it in the design of his home and studio at Kings Road, directly across the street from one of Gill's most significant residential projects, the Dodge House in Los Angeles. Schindler's use of the method in his Kings Road house was, at best, secondary, but still innovative. He poured one-story walls that would contain the main rooms of the connector between wings of the house on the slab, and utilized a block and pulley system to tilt them into place vertically. The edge of each wall, at the joint between panels, was left as a gap into which a thin glass "connector" articulated each panel joint.

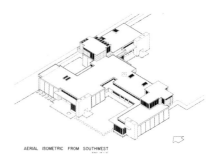

AERIAL ISOMETRIC FROM SOUTHWEST

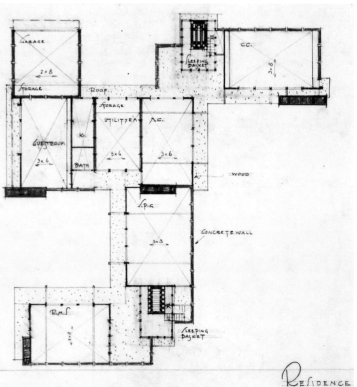

138

Steven Holl Architects—St. Ignatius Chapel

In the Jesuit's "spiritual exercises", no single method is prescribed—"different methods helped different people ..." Here a unity of differences is gathered into one. The light is sculpted by a number of different volumes emerging from the roof. Each of these irregularities aims at different qualities of light. East-facing, south-facing, west- and north-facing, all gather together for one united ceremony.

Each light volume corresponds to a part of the program of Jesuit Catholic worship. The south-facing light corresponds to the procession, a fundamental part of the mass. The city-facing north light corresponds to the Chapel of the Blessed Sacrament and to the mission of outreach to the community. The main worship space has a volume of east and west light. The concept of "different lights" is further developed in the dialectic combination of a pure colored lens and a field of reflected color within each light volume. A baffle is constructed opposite the large window of each "bottle of light." Each of the baffles is back painted in a bright color; only the reflected color can be seen from within the chapel. This colored light pulses with life when a cloud passes over the sun. Each bottle combines the reflected color with a colored lens of the complementary color. At night, which is the time of gatherings for mass in this university chapel, the light volumes shine in all directions out across the campus like colored beacons. On occasion, for those in vigilant prayer, light will shine throughout the night. The visual phenomena of complementary colors can be experienced by staring at a blue rectangle and then a white surface. One will see a yellow rectangle; this complimentarily contributes to the two-fold merging of concept and phenomena in the chapel.

The concept of "Seven Bottles of Light in a Stone Box" is expressed through the Tilt Wall method of construction. The integral color Tilt-Wall concrete slab provides a more direct and economical tectonic than stone veneer. The building's outer envelope is divided into 21 interlocking concrete panels cast flat on the chapel's floor slab and on the reflecting pond slab. Over the course of two days these panels were put in place by a hydraulic crane, which strained at the ponderous weights of up to 80,000 lbs. "Pick pockets" or hooks inset into the panels were capped with bronze covers once the panels were upright. Windows were formed as a result of the interlocking of the Tilt Wall slabs, allowing the ⅝-inch open-slab joint to be resolved in an interlocking detail.

The chapel is sited to form a new campus quadrangle green space to the north, the west, and in the future, to the east. The elongated rectangular plan is especially suited to defining campus space as well as the processional and gathering space within. Directly to the south of the chapel is a reflecting pond or "thinking field."

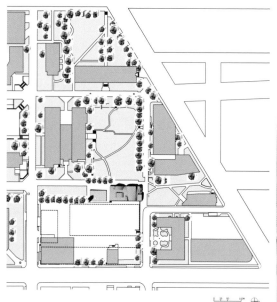

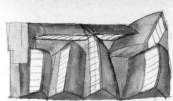

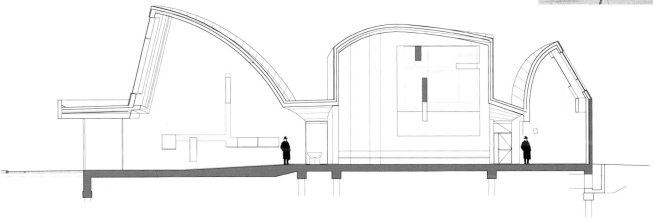

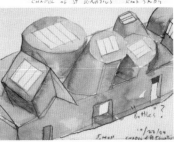

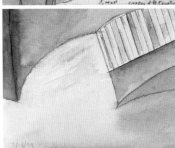

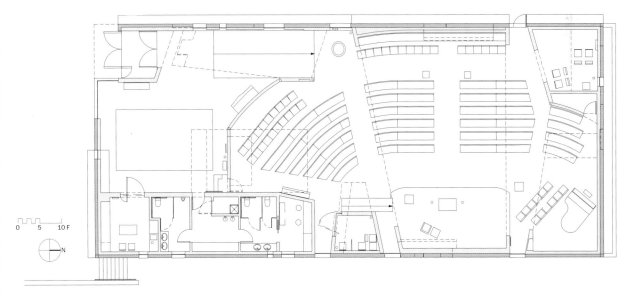

0   5   10 F

N

Cunningham Architects—TXU Customer Service Center

The Texas Utilities Customer Service Center is a 50,000-square-foot open office building constructed on a minimal budget. Located just off of a state highway, the undeveloped site offered an interesting and ideal topography to accomplish the client's goal: to be modest and sensitive to the natural surroundings. The building was nestled into the land at the far, low end of the site, blending into the rural landscape that defined the motorist's view from the highway.

As with many low-cost office buildings, the expressive qualities of the project were limited to surface treatment. Like those structures, this building is about the skin, not the immaterial skin of mirrored glass cladding, but a tangible concrete skin of stained Tilt Wall panels. While concrete tilt-up construction is inexpensive and utilitarian, its potential to be expressive is often masked by veneers. Cunningham Architects chose to exploit the material's quiet variations inherent in the process of pouring and working the concrete. The building's Tilt Wall panels were steel troweled and stained in an imprecise manner. The colors of the stain take inspiration from vines and grasses, and blend with the surroundings. The building elevation was conceived as a painterly composition, with regulated window openings, saw cuts and repetitive window openings. A line of unadorned grouted lift points punctuate across the façade completing the work.

1. BUILTUP ROOF ON METAL DECK
2. STRUCTURAL STEEL ROOF FRAMING
3. METAL WALL FRAMING W/ FULL CAVITY INSULATION
4. RECESSED SOLAR SHADE POCKET
5. THERMALLY BROKEN ALUMINUM WINDOW W/ INSULATED GLASS
6. EXPOSED HARD TROWELED TILT WALL CONCRETE PANEL W/ ACID STAIN, VARIOUS COLORS
7. REINFORCED CONCRETE FOUNDATION

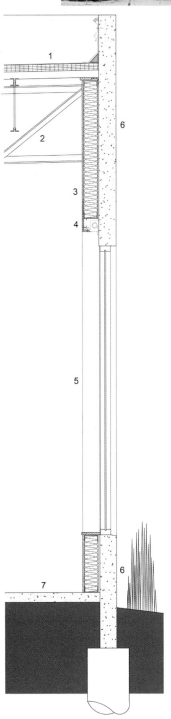

Elliott + Associates Architects—K.J. McNitt Construction Company

The owner's goal was to construct the largest possible owner-occupied office building on an existing 100 x 135 foot industrial-zoned property. The architectural concept highlights the fact that the owner is a commercial contractor with special expertise in precast concrete panel construction.

Our goal was to let the building be a living illustration of the concrete panel construction process. The honesty of panel construction is shown with various "expressions." Temporary bracing becomes permanent with 4 inch o.c. used oil field pipe structure on the exterior and interior. Panel joints are emphasized with 8-inch-wide glass slits, 4 inch o.d. oil field pipe joints and open joints. The precast concrete panel reception desk illustrates material beauty, mass (round pipe panel supports) and structural cantilevers (desk tops). Panel slits are continued through the interior spaces, which allow transparency and architectural continuity. Panel thickness (actually "thinness") is illustrated with the freestanding and vertically cantilevered west panel wall. Panel joinery is celebrated. Concrete finish options included light sandblasting (west wall) on the exterior and interior, smooth form finish (reception desk), stained concrete (floor slabs), poured-in-place plywood formed concrete (west retaining wall) and unfinished concrete (west wall west face). The parking area is designed as an outdoor "room" defined by the south building façade, the west cantilevered panels, the concrete parking area "floor" and the steel tube corner element

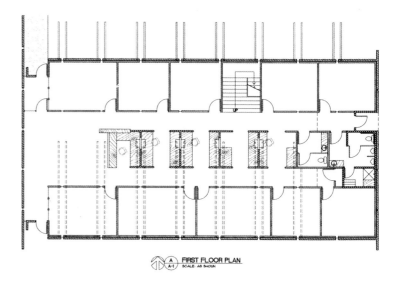

Exterior view looking south showing "freestanding" west wall panels

Exterior view looking southwest with rusted steel precast concrete supports

148

that "connects" all three dimensions of the "room" together. Metal studs and drywall panels with glass slits define interior spaces. Ceilings are exposed steel joist and deck with stained concrete and carpeted floors. This building illustrates that simple materials used in an interesting way can yield memorable results. The architecture celebrates the construction process.

Rand Elliott FAIA

West wall panels with slit view of diagonal bracing beyond

Steel-bracing detail

Steel-bracing detail and slit window

Office corridor with diagonal bracing and exposed metal studs

Interior diagonal-bracing at reception area

Second floor overlook view

Exterior view looking east with recycled stair

151

Rob Quigley—San Diego Children's Museum

This museum is part of a series of structures we have built over the years that explore Tilt Wall's potential to create civic architecture. Like the art-oriented activities within, the 50,000-square-foot San Diego Children's Museum seeks to engage and educate the users. The warm but unadorned concrete gives the project a solid, timeless, muscular, static feel, while the Tilt Wall construction method (some panels are 52 feet high and 160,000 lbs) allows for light, airy, delicate and kinetic spaces.

The Tilt Wall method allows the architecture to tell the story of its own construction. Exposed "temporary" braces holding the panels illustrate and teach the children concepts of gravity and structure (as well as deal with a tight budget). All connections and supports are exposed including the pick points.

A central glass chimney contains a transparent elevator and distinguishes the building from the surrounding commercial and residential properties. Importantly, the chimney is also a cooling tower, which exhausts hot air from the naturally cooled and heated gallery spaces. In fact, mechanical HVAC is completely absent in the major gallery spaces. Even in San Diego's benign climate, this would have been impossible without the thermal mass and surface area provided by the Tilt Wall construction. The majority of the gallery and staff spaces are naturally daylit, augmented by rooftop photovoltaic panels.

Rob Quigley FAIA

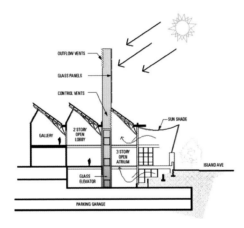

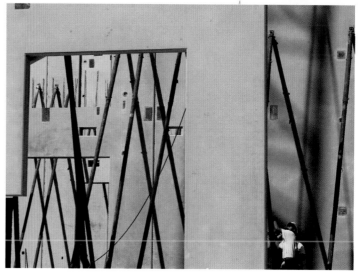

Carlos Jiménez—Rice University Library Archive

Early in the design process we chose Tilt Wall construction as the optimal system for building Rice University's Library Service Center, the inaugural structure sited in a new campus annex that we also designed for the university. What initially appealed to us was the monolithic qualities of Tilt Wall construction, as well as its efficient assembly and modest beauty. It also allowed us to explore and shape a custom profile for all exterior wall panels, animating the severity of what is technically a large mass or a vault-like refrigerant (all stored books and documents must be kept at a constant 50 degrees Fahrenheit). Casting this singular profile as integral to each wall section augmented the thickness of the entire building perimeter, providing further solidity, security and strength for the 18,500-square-foot book depository. Once all of the wall panels were assembled together (at the time the tallest Tilt Wall walls ever poured and lifted in the region), the building ignited as its rhythmic cascading pattern became a changing play of light and shadow. Sealed and painted a bright iridescent green, the building acquires the presence of a geometric topiary across the surrounding green field.

Carlos Jiménez

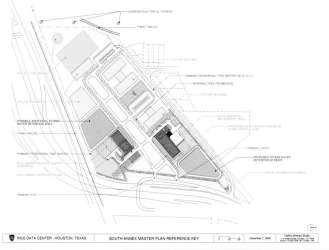

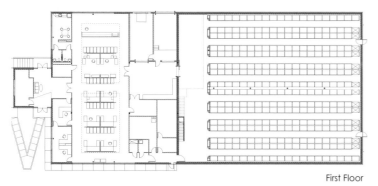

First Floor

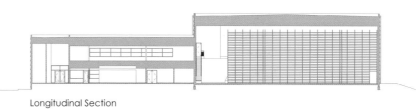

Longitudinal Section

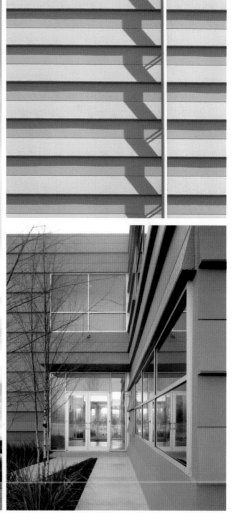

157

Patkau Architects—Gleneagles Community Center

The Gleneagles Community Centre is located on a small, gently sloping site adjacent to a public golf course. The program is organized on three levels to minimize the building footprint. By subtly reshaping the cross-sectional topography of the site, the lower level and the intermediate level are both located on grade. The intermediate level is accessible from a generous porch along the street and contains a community "living room," café, meeting room, administration, and child care facilities. The lower level opens on the opposite side of the building to the covered terraces and courtyard spaces adjacent to the golf course, and includes a gymnasium, multipurpose room, arts room, youth room, and outdoor specialty area. The upper level accommodates fitness facilities.

The sectional arrangement of interior spaces activates and energizes the building. The volume of the gymnasium rises through the three levels; walls that separate this volume from adjacent spaces are glazed to facilitate visual connection between the various programs within the building. These simultaneous views of multiple activities animate the interior; the life of the building and the energy of the place are palpable to the community within and without.

The structural system consists of cast-in-place concrete floor slabs, insulated double-wythe composite tilt-up concrete end walls and a heavy timber roof. This structure is an important component of the interior climate-control system. The structure acts as a huge thermal-storage mass, a giant static heat pump that absorbs, stores, and releases energy to create an extremely stable indoor climate, with constant temperatures inside occupied spaces, regardless of the exterior climate. Radiant heating and cooling in both floors and walls maintains a set temperature; the concrete surfaces act alternately as emitters or absorbers. The thermal energy for this system is provided by water-to-water heat pumps via a ground-source heat exchanger under the adjacent permeable parking area.

Roof overhangs provide protection from winter rains, shield interiors from excessive local solar loads in summer, and discharge rainwater into adjacent landscape swales to permeate back into the natural landscape.

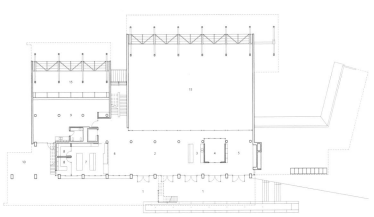

1 Entry Porch
2 "Living Room"
3 Cafe
4 Meeting Room
5 Fire Place Lounge
6 Reception
7 Administration
8 Office
9 Childcare
10 Children's Playground
11 Fitness
12 Reception
13 Training Studio
14 Counseling
15 Open to Below
16 Workshop
17 Art Studio
18 Maintenance
19 Mechanical
20 Electrical
21 Workshop Courtyard
22 Kitchen
23 Youth Lounge
24 Gymnasium
25 Storage
26 Multipurpose

Main Floor

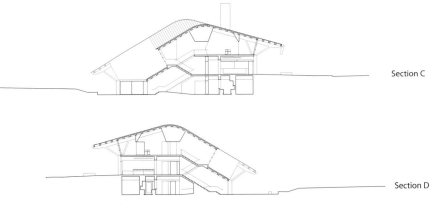

Section C

Section D

The work of Powers Brown Architecture

Our work in Tilt Wall defines only one area of our research. We like to think it benefits from this, while simultaneously contributing to the work we do in conventional structures, urban design, and interior design.

It is important to us, given all that this book covers regarding the history, workings and propositions of Tilt Wall, to put forth the following overview of how we see our practice as a whole—at once engaged in the everyday and striving to build meaningful structures, and working projectively, as put forth by George Baird and others, at the same time.

Our work, upon reflection, always leads us to consider of where we ourselves work. Our deep engagement with the architecture of commerce sharpens our focus on what we work with, how we work with it, and how it invades and invents processes. For us these questions cannot be approached without consideration of the neglected territory within which we find our efforts. This notion of neglected territory circumscribes the types of projects we are able to engage as a young practice. Later it seemed to describe the type of practice we had become. Many of our early projects, ignored by our established competitors—manufacturing facilities, big box warehouses and investment grade commercial developer projects—make up most of the built environment, but resist the type of critical investigation that avant-garde practice deems as the standard form of culturally-meaningful manifestation. They are projects driven by capital flow, financing, real-estate formulas and expediency of construction. Any perceived non-essential speculation (such as the traditional processes of criticality) is expunged from the timeline.

So much of the territory that constitutes the grounds of architectural investigation is pre-determined before the designer gets involved. We guide much of what has been presumed without the opportunity to control it. The uncomfortable and often banished rubric of style becomes the expertise of the uncritical designer in neglected territory. This is precisely why most of the built environment lies outside the shrinking reserve of programs that have architectural potential. The debate for us becomes how to re-evaluate our "critical" projects in light of "projectivity".

Projectivity as a new modality of practice was formulated during the gestation of Powers Brown Architecture, and as it has evolved as an argument in the academy, we have engaged it on the ground heuristically, in actual everyday practice. This has meant defining what has been dubbed post avant-garde practice on the fly and in real time. Working with the perception that design operations can only evolve from the formal challenges balanced against the market conditions of circumstance, we have abandoned the critical notion of resistance to consumer society in favor of merging our processes into products. Our clients weren't interested in the old criticality anyway, and as an everyday practice struggling to overcome our allotted anonymity, to resist becoming a service was easier in some ways. Our company's ethical, binding matrix comes as a result of accepting every client rather than looking for the right ones, approaching each project on its own terms, and inventing with what we find when doing so. We live with a kind of split-personality disorder: the "projects" and the "work" as it sometimes comes down. The things that seem open to potential and those that seem closed define the difference between the everyday practice and the avant-garde. The latter category of work can contribute to the former; but it is always dismissed by the critical process, yet engaged by that becoming defined as the projective, post avant-garde or, as we often seem to categorize, the mainstream. So we read Baird, Somol, Whiting, Speaks and Hickey. We talk to brokers, contractors, engineers, developers and know they are the new clients. And we press on, discovering things like Tilt Wall construction right in front of us. And, of course, we build …

Accredo Packaging, Inc.—Regional Corporate HQ and Manufacturing Plant

This building was completed for a national manufacturer of plastic grocery bags, and located in an industrial park in suburban Houston, in Sugar Land, Texas. The program is primarily driven by the plastic extrusion tower requiring nearly 90 feet of clear space to operate. Ancillary functions include ink processing and supply, materials storage, and office support with quality control labs.

Planes both dominate and balance the larger volumes. Combined with an abstract composition used to camouflage the scale of the tower, the planes index program components and express formal conditions such as entry and fenestration. They are organized to produce a clear hierarchy yet compliment, rather than compete with, the tower. The whole becomes a dynamic composition that simultaneously expresses the content of the program and the quality of internal work space.

THE ABSTRACT COMPOSITION OF PLANES INDEX PROGRAM COMPONENTS SUCH AS OFFICES, MAIN ENTRY AND BREAKROOM

CRANE TOWER

INK-MIXING ENCLOSURE

OFFICE CLERESTORY

STORAGE & MANUFACTURING PLYNTH

FUTURE EXPANSION

SITE

CARDINAL MEADOW DR

DAIRY ASHFORD DR

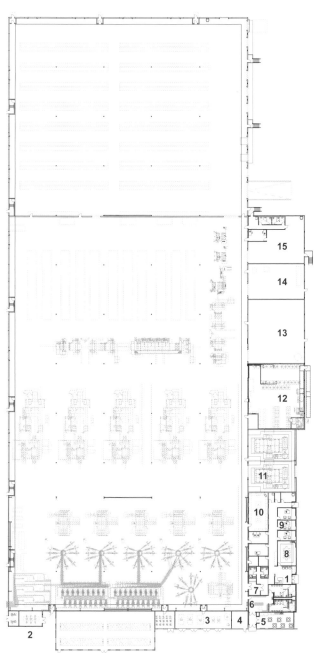

1. MAIN ENTRY / RECEPTION
2. SERVICE YARD
3. CONTINUOUS CANOPY ABOVE
4. FIRE PUMP
5. LUNCH ROOM
6. LOCKER ROOM
7. CHANGING ROOM
8. CONFERENCE
9. OFFICES
10. QUALITY CONTROL
11. OXIDIZERS
12. INK ROOM
13. CLEAN ROOM
14. PLATE ROOM
15. MAINTENANCE PARTS

A 90'-0" TALL EXTRUSION TOWER CLADDED WITH INSULATED METAL-PANEL SKIN WITH SLIT WINDOWS TO PROVIDE DAYLIGHT INTO THE FACILITY

CONSTRUCTION PROCESS: Tilt Wall construction was used to sponsor the breakdown of the slab into a volume composed of planes and planar elements.

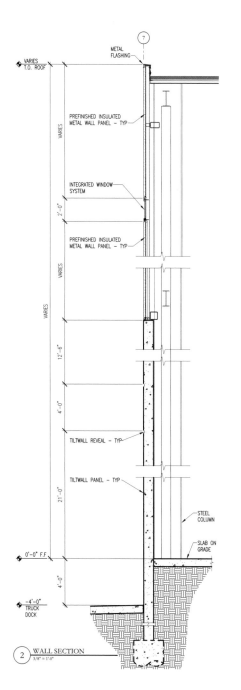

WALL SECTION

WAREHOUSE
CLERESTORY
OPENING IN
CONCRETE

OFFICE
AREA
PLANAR
ELEMENTS

168

We wanted to emphasize the main piece of equipment or at least the shell around it which, of course, had to be VERTICAL. This left a COUNTERPOINT STRATEGY of forcing the supporting program elements into a HORIZONTAL slab. The architecture of the tower element is consciously graphic, like the labels on the bags it produces, mitigating its size with an ABSTRACT PATTERN composed of textures and colors. Random windows infest the metal panel pattern to allow light to filter inside the plant, serving as an interest-holding INTERNAL LIGHTSCAPE animating the mundane task of monitoring the operation of the machine.

HORIZONTAL VS. VERTICAL METAL PANEL COMPOSITION

Kelsey Seybold—Corporate Offices

This 170,000 square foot administrative office building sits on 16 acres in Pearland just south of downtown Houston, Texas. The site is unique in that it sits outside of an urban context, and also outside of a banal, suburban, office-park influence—it is nestled alongside a creek watershed and a retention lake. The site was surrounded by these two indeterminate conditions, but inside the site, we speculated something quite different was happening. The client was interested in both the site and a typical building floor plate due to economic pressures—the site was not typical and thus affordable, while the buildings needed to be typical and thus efficient.

We combined the two antagonistic systems—the sites irregularity at the creek edge and the typicality of floor-plate size based upon program—to create an efficient yet inflected and agitated finial plan form. The typical "slipped Z" suburban floor plate, in its ideal condition, can be both efficiently planned and built; we wanted to distort it or contextualize it, yet minimize the impact on these qualities. Ultimately we were able to achieve a balance of the 26 independent and non-Cartesian planes that comprise the final volume by utilizing Tilt Wall construction to maintain cost efficiency and provide formal flexibility.

The floor plates are approximately 50 thousand square feet for the first three levels. Each level contains a "town-square" area located in the round drum facing the landscaped park between the creek edge and the building. The town squares are informal eating, coffee and private Wi-Fi areas for employees. Core-area conference rooms are also found at each level, with a grand creek-side cafeteria and training suites on the ground floor. The fourth floor contains executive departments with rooftop terraces.

Our goal was to use the thematic system of exterior brakes and inflections related to the topography in order to influence "way-finding" and a sense of location in the interior—issues that are often unaddressed in open-office environments. We developed simple but sophisticated tripartite color / material / texture systems to indicate both "creek-side" and "away-side" interior spaces, but also those from level to level, incrementalizing the building into digestible and comfortable subdivisions and departments that resonate with the exterior form.

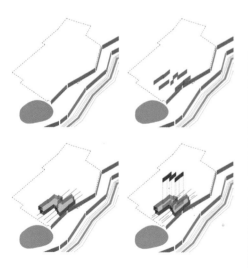

172

173

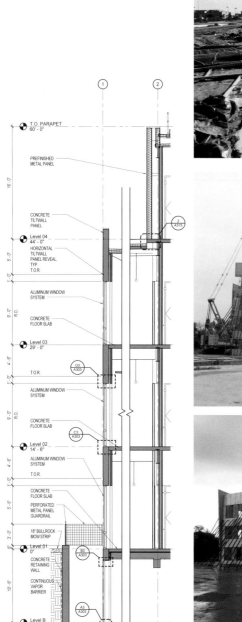

SECTION THRU METAL PANEL / WINDOW / TILTWALL 2

T.O. PARAPET
60' - 0"

PREFINISHED
METAL PANEL

CONCRETE
TILTWALL
PANEL

2
A315

Level 04
44' - 0"

HORIZONTAL
TILTWALL
PANEL REVEAL
TYP
T.O.R.

ALUMINUM WINDOW
SYSTEM

CONCRETE
FLOOR SLAB

Level 03
29' - 0"

T.O.R.

D3
A303

ALUMINUM WINDOW
SYSTEM

CONCRETE
FLOOR SLAB

C3
A303

Level 02
14' - 6"

ALUMINUM WINDOW
SYSTEM

T.O.R.

CONCRETE
FLOOR SLAB

PERFORATED
METAL PANEL
GUARDRAIL

18" BULLROCK
MOW STRIP

B3
A303

Level 01
0"

CONCRETE
RETAINING
WALL

CONTINUOUS
VAPOR
BARRIER

A3
A303

Level B
-10' - 6"

A4 SECTION THRU METAL PANEL / WINDOW / TILTWALL 2
SCALE: 1/4" = 1'-0"

Eagle Burgmann—USA Headquarters

This project, for a German holding company's American division, had the simple objective of capitalizing on its proximity to a major, well-traveled highway to increase its presence and brand. The existing facility was a functional single-story industrial box that facilitated the manufacturing and testing of valves. The consolidation of the off-site management arm of the company, combined with office space for a recent acquisition, had to be seamlessly combined to form the new headquarters.

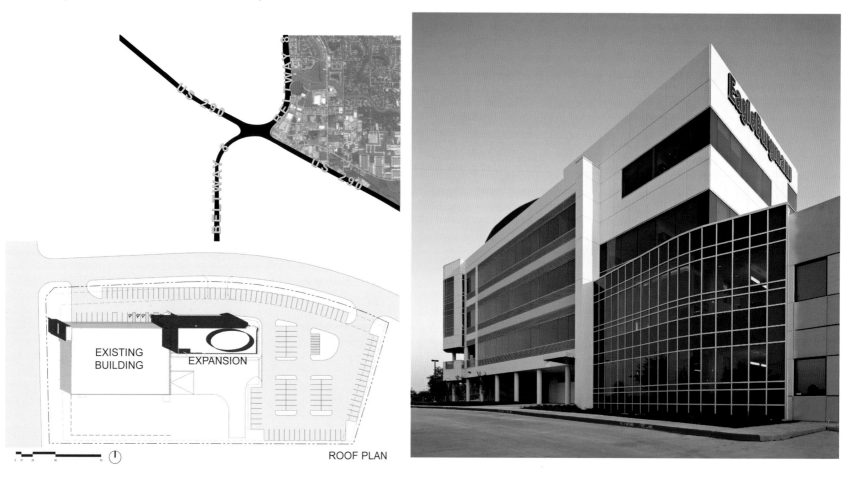

EXISTING BUILDING

EXPANSION

ROOF PLAN

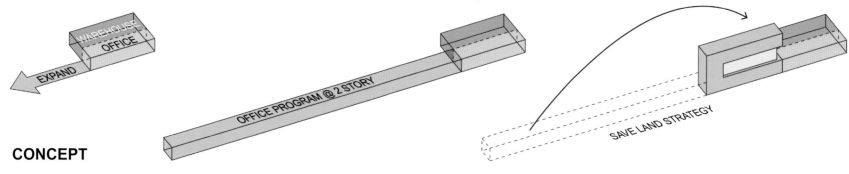

WAREHOUSE

OFFICE

EXPAND

OFFICE PROGRAM @ 2 STORY

SAVE LAND STRATEGY

CONCEPT

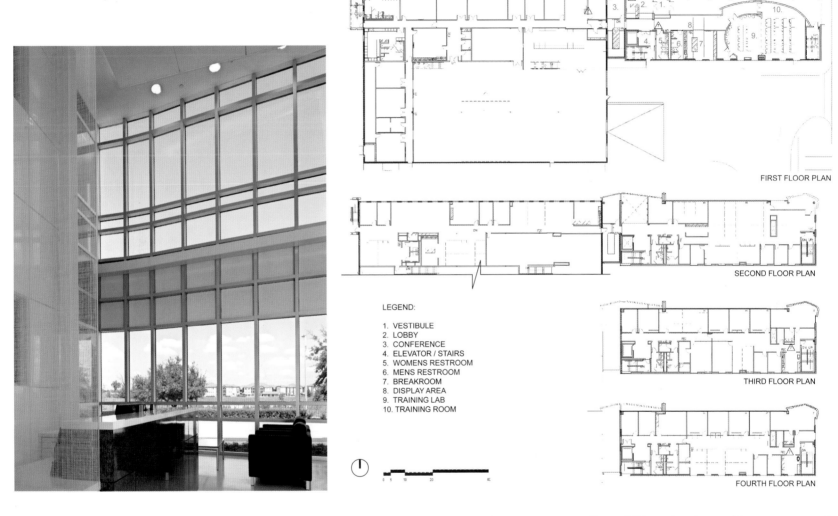

FIRST FLOOR PLAN

SECOND FLOOR PLAN

LEGEND:

1. VESTIBULE
2. LOBBY
3. CONFERENCE
4. ELEVATOR / STAIRS
5. WOMENS RESTROOM
6. MENS RESTROOM
7. BREAKROOM
8. DISPLAY AREA
9. TRAINING LAB
10. TRAINING ROOM

THIRD FLOOR PLAN

FOURTH FLOOR PLAN

0 5 10   20   40

The scheme strategy was driven by the limitations of the site—while large enough for the two-story office addition, the site could not accommodate the required parking. Consequently, a vertical building allowed for the iconographic expression and the flexibility to seam into the mezzanine level of the factory. As the original horizontal building had been made using Tilt Wall construction, we chose to use it for the relatively compact vertical building in an effort to insert a constant into a problem that seemed to be all variables.

The lower levels of the addition are training and vendor spaces, with factory operations and engineering at the second level spanning both the new and existing footprints. Above, in the new tower, are executive spaces for both divisions with concomitant support spaces. The tower is integrated with the factory building by a kind of vertical / horizontal transaction zone—a knot of circulation that allows the floor-to-floor heights at levels one and two to calibrate through soft ramps.

179

Concentrating the vertical tower circulation in the same zone and making this the facility entry creates an active and vital terminal-like space always populated and buzzing while dispersing employees out to the factory or up to the office areas. From this core, the break rooms and gathering spaces are layered to feed off of the activity and buffer the less kinetic program elements.

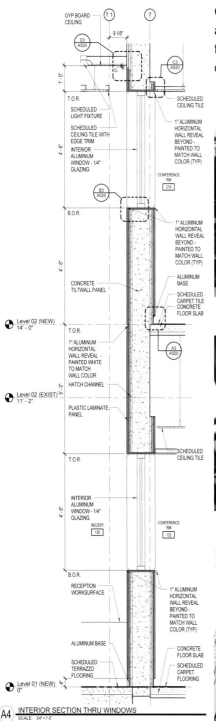

GYP BOARD CEILING

9 1/8"

D3
A520

C3
A520

T.O.R.
SCHEDULED LIGHT FIXTURE
SCHEDULED CEILING TILE WITH EDGE TRIM
INTERIOR ALUMINUM WINDOW - 1/4" GLAZING

SCHEDULED CEILING TILE
1" ALUMINUM HORIZONTAL WALL REVEAL BEYOND - PAINTED TO MATCH WALL COLOR (TYP)

CONFERENCE RM 219

B3
A520

B.O.R.

1" ALUMINUM HORIZONTAL WALL REVEAL BEYOND - PAINTED TO MATCH WALL COLOR (TYP)

CONCRETE TILTWALL PANEL

ALUMINUM BASE
SCHEDULED CARPET TILE CONCRETE FLOOR SLAB

Level 02 (NEW)
14' - 0"

T.O.R.
1" ALUMINUM HORIZONTAL WALL REVEAL - PAINTED WHITE TO MATCH WALL COLOR
HATCH CHANNEL

A3
A520

PLASTIC LAMINATE PANEL

Level 02 (EXIST)
11' - 2"

SCHEDULED CEILING TILE

T.O.R.

INTERIOR ALUMINUM WINDOW - 1/4" GLAZING

RECEPT 135

CONFERENCE RM 132

B.O.R.
RECEPTION WORKSURFACE

1" ALUMINUM HORIZONTAL WALL REVEAL BEYOND - PAINTED TO MATCH WALL COLOR (TYP)

ALUMINUM BASE

SCHEDULED TERRAZZO FLOORING

CONCRETE FLOOR SLAB
SCHEDULED CARPET FLOORING

Level 01 (NEW)
0"

A4 INTERIOR SECTION THRU WINDOWS
SCALE: 3/4" = 1'-0"

The skin of the tower is designed with solar orientation in mind. The north-facing façade, behind which the open office areas are located, is clad in high performance curtain wall while the direct west-facing elevation is punctuated with punched openings that directly index the proportions on the factory building. Most of the closed office spaces reside on this side. Each of the upper-office tower levels has a curved balcony providing outside spaces for smoking or private calls.

Texas Steel—Manufacturing Offices

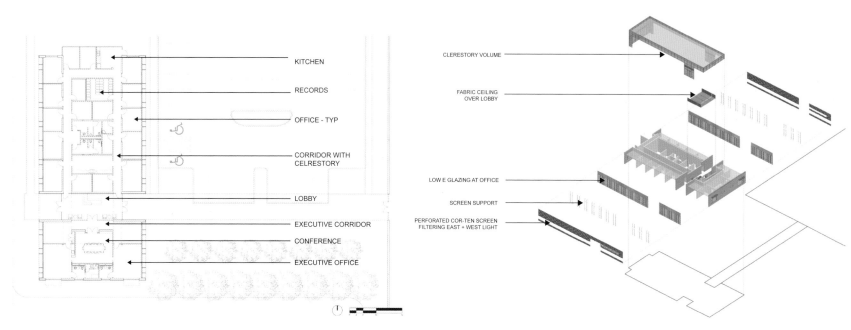

KITCHEN

RECORDS

OFFICE - TYP

CORRIDOR WITH
CELRESTORY

LOBBY

EXECUTIVE CORRIDOR

CONFERENCE

EXECUTIVE OFFICE

CLERESTORY VOLUME

FABRIC CEILING
OVER LOBBY

LOW E GLAZING AT OFFICE

SCREEN SUPPORT

PERFORATED COR-TEN SCREEN
FILTERING EAST + WEST LIGHT

The project is the addition of a small office building for the executive and support personnel for TEXAS STEEL PROCESSING, a carbon alloy and stainless steel plate-processing facility in Houston that specializes in precision plasma cutting, rolling and press breaking with CNC technology.

Weathering-steel screens, indexical to the program, allowed the building to perform auto-didactically. Combining the logic of machining and digital input, we came up with a pattern, and the client, who has never before provided services for architectural applications, manufactured them.

The preconceived economically driven solution was to include this office space as an expansion to the existing 200,000 square foot manufacturing facility. In other words "just make it bigger". We conducted a site analysis revealing a discordant juxtaposition between the scale and function of the big-box building with the residential neighborhood it is sited within. Separating the office functions into a smaller building contextually compatible with existing building patterns was an eminently more sensitive response to the problem. This solution provided a campus-like atmosphere as well as a new building with daylight on all four sides of its perimeter.

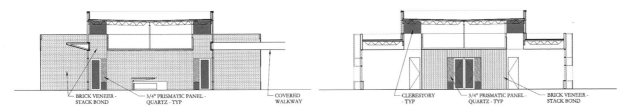

BRICK VENEER - STACK BOND — 3/4" PRISMATIC PANEL - QUARTZ - TYP — COVERED WALKWAY — CLERESTORY - TYP — 3/4" PRISMATIC PANEL - QUARTZ - TYP — BRICK VENEER - STACK BOND

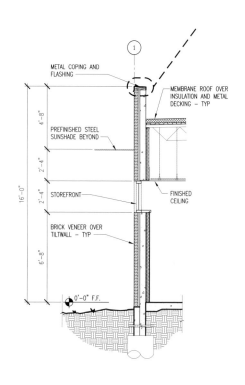

METAL COPING AND
FLASHING

MEMBRANE ROOF OVER
INSULATION AND METAL
DECKING – TYP

PREFINISHED STEEL
SUNSHADE BEYOND

STOREFRONT

FINISHED
CEILING

BRICK VENEER OVER
TILTWALL – TYP

4'-8"

2'-4"

2'-4"

6'-8"

16'-0"

0'-0" F.F.

2 WALL SECTION AT HIGH STOREFRONT
3/8" = 1'-0"

A		24 PANELS
B		16 PANELS
C		24 PANELS
D		16 PANELS
E		6 PANELS
F		4 PANELS
G		12 PANELS
H		8 PANELS
I		2 PANELS
J		2 PANELS
K		6 PANELS
L		4 PANELS

0 2 4   8   16

The cross section reveals a building within a building. A clearly defined four-sided glass box gives the building its office iconography while allowing natural light into the public corridors. The lower building is a tectonic expression with brick planes running east-west indexing the residential character of the neighborhood, and user-generated steel screens running north-south diffusing the abundance of natural light allowed into the perimeter office areas. We utilized Tilt Wall panels as structural or infrastructural planes to form the interior space while economically providing the height to allow the clerestories to work.

During the process of making this building, in a collaborative effort with our client, we explored the intersection of pattern and composition, working toward the compounded argument of formal validity and economic viability, while moving beyond the limitations of repetitive industrial production. The idea was to give the building an iconographic presence that supported its purpose without recourse to untransformed literal devices.

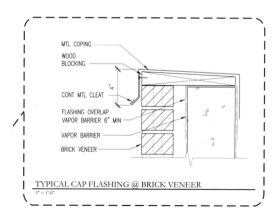

MTL COPING
WOOD BLOCKING
CONT MTL CLEAT
FLASHING OVERLAP
VAPOR BARRIER 6" MIN
VAPOR BARRIER
BRICK VENEER

TYPICAL CAP FLASHING @ BRICK VENEER
3" = 1'-0"

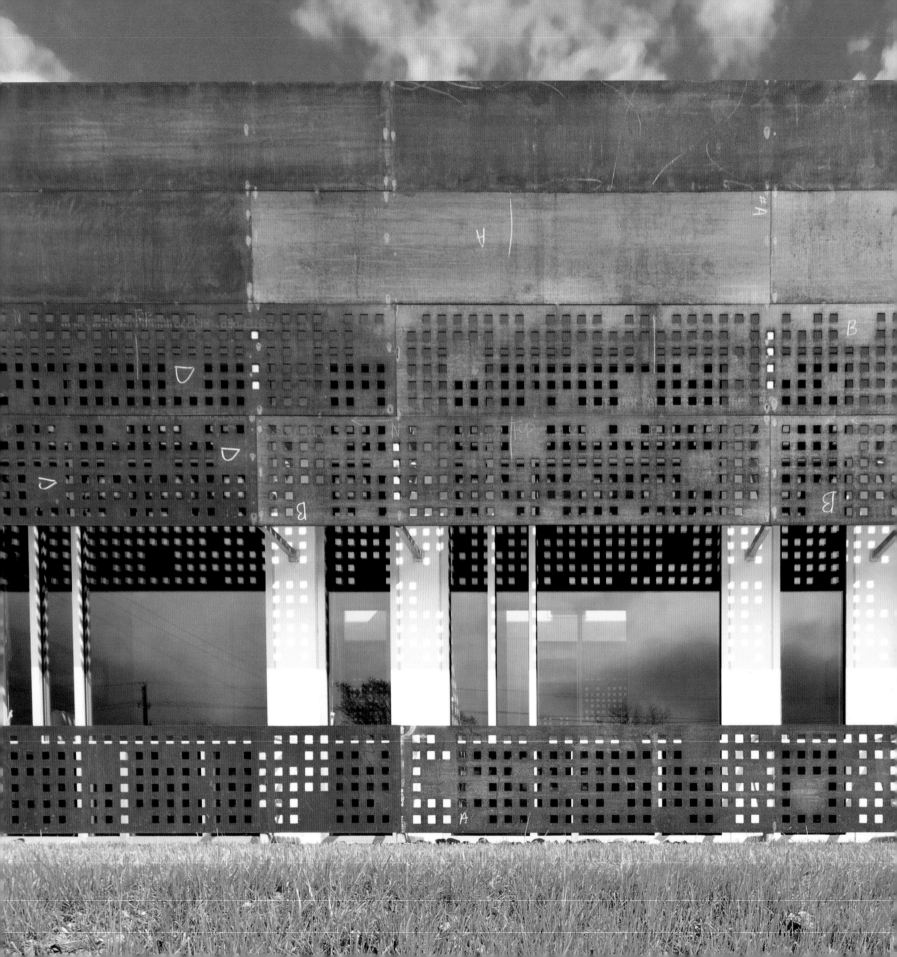

Fort Bend County War Veterans Memorial

This monument is offered to the veterans of Fort Bend County as an inadequate acknowledgement of their service to the United States of America. Interpreting the monument requires a willingness to accept it that appeals to many; it must create a sense of both ambiguity and clarity. The Sugar Land Memorial Park combines the contemplation of war and the celebration of veterans into a single element. It provides an experience that weighs the costs of war with the rewards of potential victory. It was built as part of an organization's donation to the city, and all services that contributed to its completion were pro-bono. It was conceived utilizing Tilt Wall construction as a demonstration of the potential of the technique.

The most important aspect of a memorial is to honor veterans. The experience is established by two major architectural components set in juxtaposition—one celebratory, the other contemplative. The power of the monument, its ability to evoke a meaningful effect on its visitors (both veterans and their supporters) is driven by the strong contrast between these two concepts.

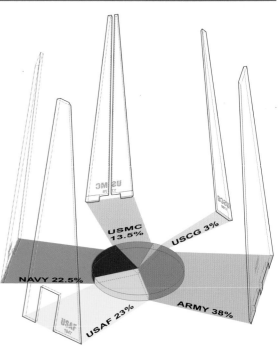

The "Flag Panel"

The celebratory aspect of the monument experience is embodied in the Flag Panel. It relies upon literal and referential symbolism. There is an abstract but identifiable flag carefully accented with blue tile and red concrete stain. The panel itself is flag shaped and utilizes a structural flag panel to allow a cantilever, which creates an open gateway effect that is anchored in the lake. The intent is to create an emotional effect, crossing from the profane to the sacred. The flag panel acts as a gateway to a symbolic river and is designed to evoke a sense of pride, a sense of the physical manifestation of service, duty to country, and survival.

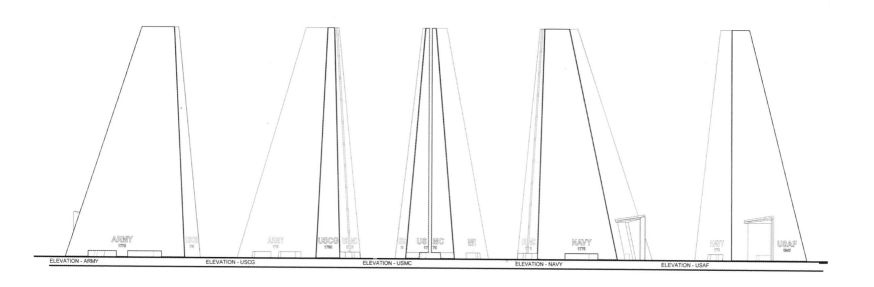

ELEVATION - ARMY   ELEVATION - USCG   ELEVATION - USMC   ELEVATION - NAVY   ELEVATION - USAF

The "Remembrance Tower"

The tower that is placed directly in the path of the Flag Panel gateway is deliberately ambiguous in form, contrasting with the simple analogy of the Flag Panel. Its verticality also counter poses the horizontal boundary-defining gateway of the Flag Panel. Its basic form is both familiar and obelisk-like, and at the same time hard to identify. The tower is comprised of wall panels huddled together on the island site. They sit within an equal-sided pentagon form. The five branches of the military are symbolized by the pentagon shape and the five huddled wall panels. Each panel is 50 feet tall (10 feet for each branch), and the base width of each of the panels is driven by the relative size of enrollment in each branch. Thus the Army panel is the widest at the bottom. They all resolve to similar widths at the top where they form an oculus.

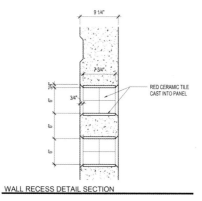

BRIDGE WALL JOINT DETAIL

TILT WALL PANEL ACUTE JOINT

WALL RECESS DETAIL SECTION

As it is meant to evoke a visceral sense of remembrance, the form and materials are activated by contrasting sunlight. The entry into the tower symbolically faces west, where the day ends, and allows the visitor to face east towards a new beginning. Inside a tapered view of the sky anchors the eastern-oriented light slot. Throughout the day a sunbeam traverses over the internal faces of the panels, awakening with light the names of the fallen inscribed on the internal panel faces. The tactile senses are engaged as the rustication at the base of the internal patterns progresses from rough to smooth symbolizing the progress from chaos to calm.

The monument balances the accessible imagery of the Flag Panel with the emotive and ineffable experience inside the Remembrance Tower. In so doing it means what it needs to mean to a diverse constituency of individuals who made the sacrifice to serve, and who come to recall those that paid the ultimate price for doing so.

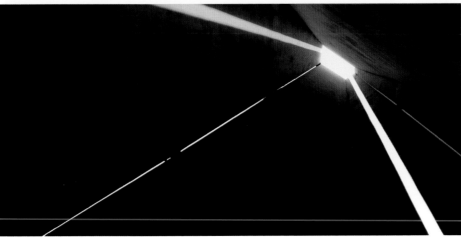

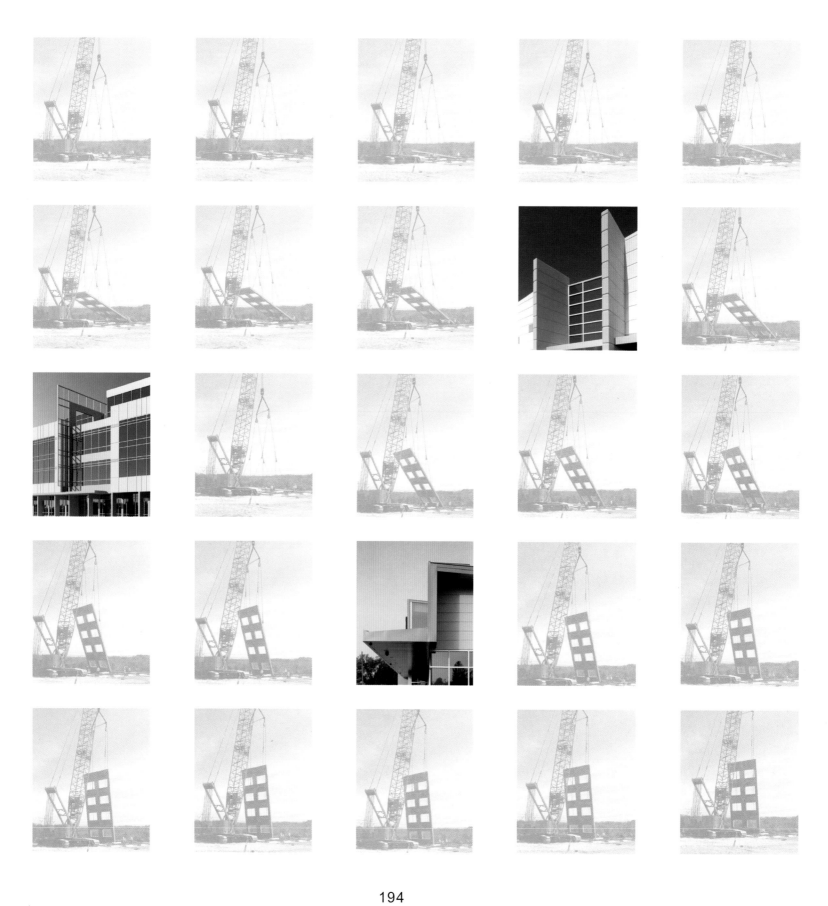

Small Smart Boxes

The next big thing in Tilt Wall may actually be small. While nearly every building type has now been executed in Tilt Wall, this section focuses on recent projects we have bundled together as "Small Smart Boxes". These projects are from a series of small corporate build-to-suits that, because they had a portion of warehouse, manufacturing or distribution in their programs, and were pre-determined to be built using Tilt Wall. This series of buildings demonstrates the architectural potential of this technology— its ability to provide interesting and formally innovative work environments / corporate image projections in a low cost form.

By way of a back story to this declaration, well known avant-garde architect Peter Eisenman once said that any hack can build a dumb box. Assuming he was elaborating on what has become known as the "big dumb box", several conjectures are inspired by his admonition. We know that after Tilt Wall construction's initial flirtation with high Modernism (first through Gill and in the 20s via Rudolph Schindler in the Kings Road House) that the technology was relegated to design anonymity until after World War II when it became associated with big-box warehouse and post-war distribution building architecture. Somewhere along the line, and despite the innovation of Albert Kahn and others in industrial architecture, the big boxes became, well, dumb. Devoid of detail and calibrated for scale by way of the economics of speed of construction and efficiency of scale, Tilt Wall seemed immune to the ongoing experimentation in form making, and the innovative utilization for the creation of space. There was a general sense that unless it was a box, the technology just didn't have much to offer.

But all boxes aren't the same. Small Smart Boxes is a line of buildings backed by an approach to solving the problem of our day post Great Recession—will design innovation suffer due to low budgets? As the market returns, architects, engineers and our consultants have begun to understand the cliché "New Normal". Traditional development for both commercial developers and corporate end-users is now subject to increased financing scrutiny, a huge increase in equity and complex financing requirements meant to assuage another building bubble. In many ways these structural changes in the finance of architecture are suppressing the market for those that don't have cash or secure banking relationships. However, close examination of statistics and trends revealed a business opportunity for us. Basic research into the average age of small-business CEO's shows interest rates continue to be at historic lows and most small businesses have less complicated banking relationships. For us that conjured an opportunity, at a basic level, which involved business speculation intersecting design needs—how many of these CEO's looking at retirement in a ten- to fifteen-year window were interested in capitalizing on the low interest rates to build a legacy building rather than one-last lease renewal? And how many wanted a low-cost but high-quality work environment for their employees? It turns out, quite a few.

By adopting a sense that clients really no longer exist, only markets, we realized that markets require commodities. We put together a product we called Small Smart Boxes. It combined our traditional architectural services such as

programming and innovative form making with our experience in the low cost technology of Tilt Wall construction. We targeted 30- to 50-thousand-square-foot office-warehouse clients as our back up research showed us that we could generate a slab big enough for casting panels but small enough to control the building envelope architecturally (or at least we speculated the scale to be big enough to balance the formal and the efficient). We created a few case study projects, built cost models around them with the help of collaborating contractors, and went to the market in late 2004. Our upfront work in getting our product line established was based upon our ability to save in most cases between $7 to $10 per square foot over conventional construction for similar configurations.

Thirty buildings (and counting) later, we are comfortable labeling this series of buildings Small Smart Boxes. That they are small is a relative claim supported by the markets we have executed them in; none exceed 60 thousand square feet. That they are smart is a proposition that only the users and perhaps observers can authenticate. Most of the buildings in this series use Tilt Wall as a load-bearing exterior structure that is then formally attacked and eroded to defeat the box effect. In some cases we have created voluminous walls—deep-set windows or layers of penetration that defeat the thinness of the walls. In others we have agitated the panel configuration by torqueing to instigate non-Cartesian plans, or shearing to create roof lights and clerestories economically. There are endless possibilities for assembly strategies—the series of buildings included here represent sample accumulations of the incremental explorations of Tilt Wall's non-technical formal capabilities.

Finally they are boxes—only as a reminder that resisting dumbness is the best hope for revealing the potential of Tilt Wall architecture. We want these projects to instigate the possibility of un-tethering Tilt Wall from its presumed teleological circumscription as "structure".

Many of our projects, like Small Smart Boxes, make up most of the built environment but resist the type of critical investigation that leading practice implies. They are projects driven by capital flow, financing, real estate formulas and expediency of construction, such that any perceived non-essential speculation is expunged from the timeline. The owners of small businesses (that supply many of the opportunities we encounter) are interested in sponsoring environments that exceed the mundane, but within the limitations of the burden their business can carry. Should the low budget / small business client that employs the bulk of workers in the country be excluded from the benefits of high architecture? Not if they employ the Small Smart Box strategy.

Laversab

From this modern facility, Laversab will command a worldwide channel of sales representatives and distributors reaching a wide variety of scientific, government and industrial markets, including a current customer base of over 50 Fortune 500 companies As a lead supplier of industrial computers servicing the government, aviation and oil-field industries, this facility is designed to reflect the company's ongoing pursuit to compete and be the world leader in their respective markets.

But it is located in a rear guard back site in an industrial business park. And the program of just 25,000 square feet makes it one of the smallest buildings in the park. And they were interested in image …

Since the company is new, an extensive programming exercise was completed looking at growth over two, three and five years. The owner chose to utilize the five year plan. The design challenge was to mediate between the high-tech nature of Laversab's company with the context of the typical office / warehouse buildings ubiquitous to the industrial park it was sited in. The small program, end-of-the-street-site problem has a very famous precedent and we took heed of its lesson. At the Villa Besnus in Vaucresson, France, in 1922, Le Corbusier was faced with a similar problem: the need for scale with very little program. He famously turned the stair parallel to the main façade in its own volume, doubling the surface area of that main elevation. In our case we created a single, tall, curved, floating Tilt Wall plane following the gentle bend in the street to increase the apparent scale of the building.

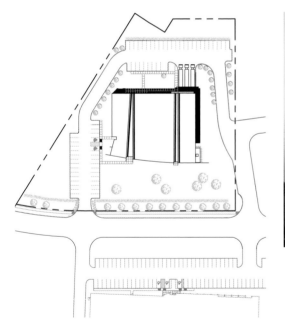

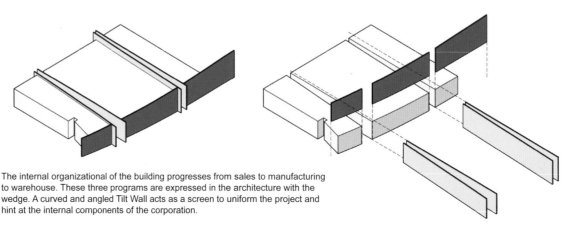

The internal organizational of the building progresses from sales to manufacturing to warehouse. These three programs are expressed in the architecture with the wedge. A curved and angled Tilt Wall acts as a screen to uniform the project and hint at the internal components of the corporation.

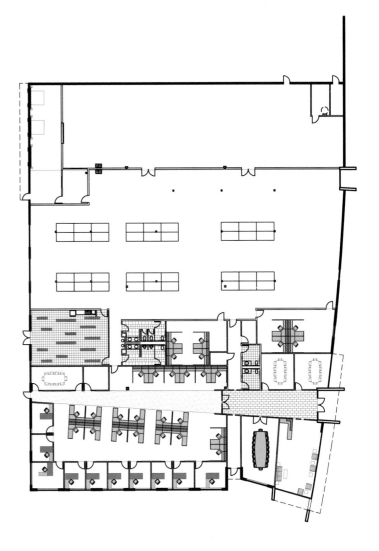

The overriding compositional strategy became a concrete screen wall with the internal building programs attached behind it. Rather than fully masking the building from the street, we allowed the screen wall to be affected by the functions behind it—vertical fins with horizontal reveals that continue through the main corridor of the office space to announce the entrance. The program is zoned into three zones: perpendicular to the main skin, office, and manufacturing and storage / warehouse.

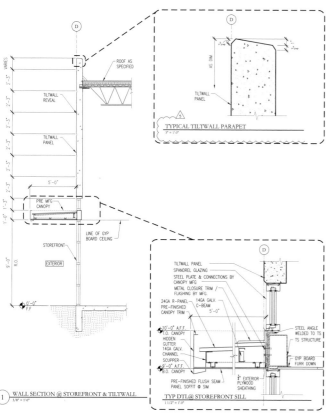

Curtain wall with an aluminum sunshade slips through the screen at the building's public areas and an abstract set of apertures punched into the horizontal reveal the pattern index and the kinetic character of the manufacturing spaces. Adjusting to the contour of the street frontage, the slight curve in the screen wall allows the building to seat itself into the site.

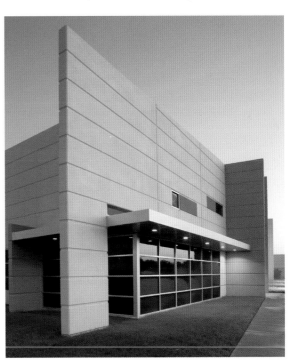

Caterpillar

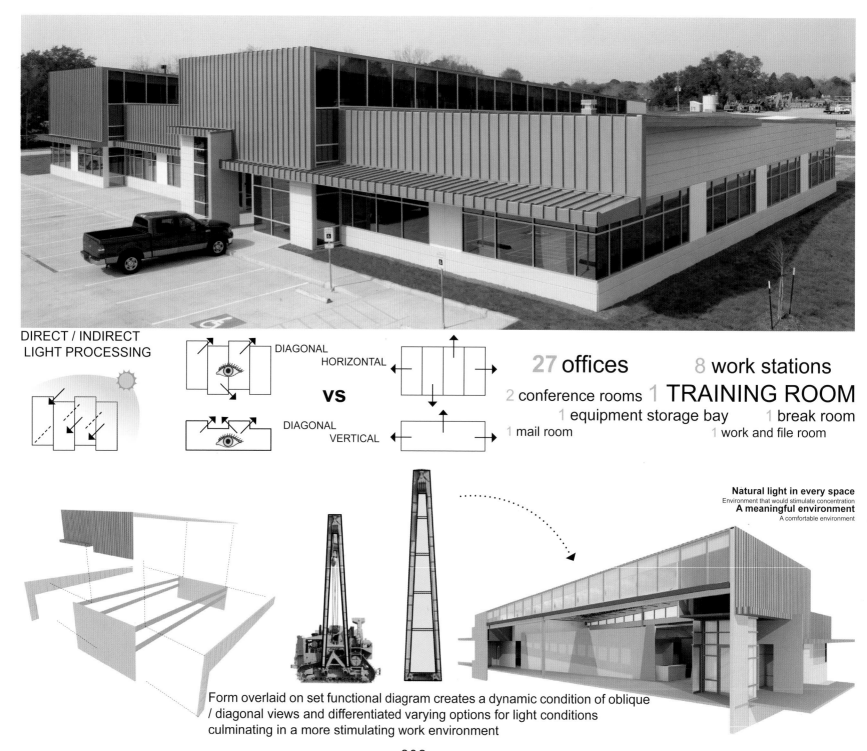

DIRECT / INDIRECT
LIGHT PROCESSING

DIAGONAL
HORIZONTAL

VS

DIAGONAL
VERTICAL

27 offices   8 work stations

2 conference rooms 1 TRAINING ROOM

1 equipment storage bay   1 break room

1 mail room   1 work and file room

Natural light in every space
Environment that would stimulate concentration
A meaningful environment
A comfortable environment

Form overlaid on set functional diagram creates a dynamic condition of oblique
/ diagonal views and differentiated varying options for light conditions
culminating in a more stimulating work environment

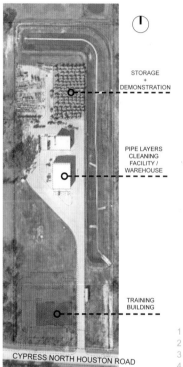

STORAGE + DEMONSTRATION

PIPE LAYERS CLEANING FACILITY / WAREHOUSE

TRAINING BUILDING

CYPRESS NORTH HOUSTON ROAD

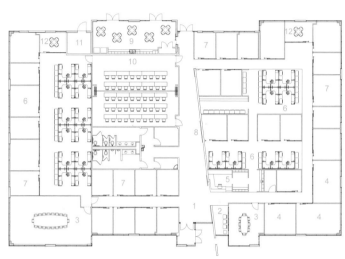

1	Lobby	5	Receptionist	9	Break Room
2	Waiting Area	6	Opened Office	10	Training Room
3	Conference Room(s)	7	Closed Office(s)	11	Mechanical Room
4	Executive Office(s)	8	Feature Wall	12	Huddle Room(s)

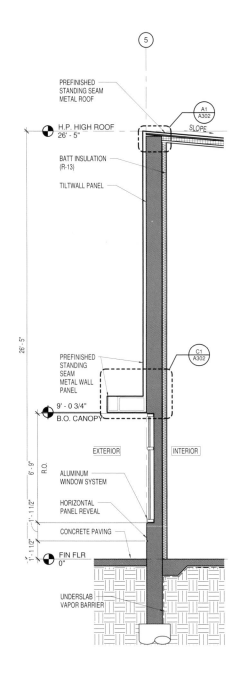

PREFINISHED
STANDING SEAM
METAL ROOF

A1
A302

H.P. HIGH ROOF
26' - 5"

SLOPE

BATT INSULATION
(R-13)

TILTWALL PANEL

26' - 5"

PREFINISHED
STANDING
SEAM
METAL WALL
PANEL

C1
A302

9' - 0 3/4"

B.O. CANOPY

EXTERIOR   INTERIOR

ALUMINUM
WINDOW SYSTEM

6' - 9"

R.O.

HORIZONTAL
PANEL REVEAL

1'- 11/2"

CONCRETE PAVING

1'- 11/2"

FIN FLR
0"

UNDERSLAB
VAPOR BARRIER

A3 WALL SECTION THRU HIGH ROOF @ PANEL LEG
SCALE: 3/8" = 1'-0"

Colt

Colt International is a fast-growing company that facilitates executive air travel on private jets across borders and nationally. The company outgrew its start-up space and purchased an economical infill site in an older office park between Downtown Houston and the Gulf of Mexico. Colt is run by two owners, each emphasizing a particular aspect of the business. Malcolm oversees the area of the business that monitors flights; he plans routes and tracks the progress of the planes. Joel leads the firm's general support, including applications for permission to land, passport / visa issues, and fuel purchase agreements. The organization of the building directly indexes this structure.

The two rooms holding the collective aspects of each of these programs are formed by a "S" shaped bar that provides general support. Each room is sized according to the number of employees within it (flight monitoring being the smaller group), and both are lit accordingly. Each also has appropriate exterior access—the park being in the coastal plain and still in possession of typical conifer trees. The two "zones" are created with a Tilt Wall snake form that becomes both wrapper and divider. The view is accommodated by an oversized picture window. Each of these picture windows is then screened by a canvas awning distorted to evoke the skin / frame of early planes. These awnings in turn create an outside interstitial porch landscape for employee use.

This overall arrangement provides several "routes" to occur in the project, the most important being the prospective client tour. Beginning in the lobby, visitors travel through a small passage to the conference room, then directly out to an elevated dais overlooking the bank of flight monitoring screens in the small room. Other routes allow for a direct interface between the two distinct functions of the business. Employees can short circuit the two by a series of "cut-throughs" that allow for quick coordination of a route in progress with fuel-purchase paperwork.

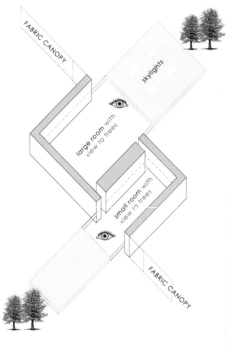

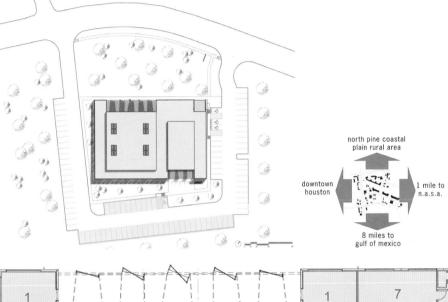

north pine coastal
plain rural area

downtown
houston

1 mile to
n.a.s.a.

8 miles to
gulf of mexico

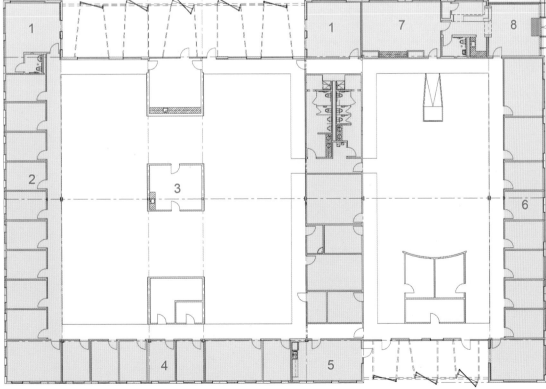

1. Executive Offices
2. Sales Department
3. Work Room
4. Accounting Department
5. Break Room
6. Flight Department
7. Conference Room
8. Entry Lobby

207

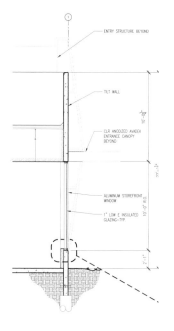

ENTRY STRUCTURE BEYOND

TILT WALL

CLR ANODIZED AVADEK
ENTRANCE CANOPY
BEYOND

ALUMINUM STOREFRONT
WINDOW

1" LOW E INSULATED
GLAZING—TYP.

Casimir

The invention in this project came from leveraging two catalysts: 1) the perpendicular street orientation of a conventional parallel orientation for a strip shopping center and 2) the gerrymandered "cross parking" arrangement that allowed for differing densities across the daily hours of use.

The perpendicular placement of the building defeated another of the "typical" approaches to hierarchy in strip center design: center emphasis and end cap articulation. A close study of the parking agreement revealed that the document was imperfectly amended several times over a number of leases across almost 20 years. We agreed the solution was to almost double the number of parking spots—available at the front third of the site—required by code for retail use after three o'clock on weekdays and noon on weekends. This supported the idea of doubling the head portion of the building mass, essentially adding second story, and this, in turn, invited the opportunity to look to alternate programs. The tail portion is formally animated with a kinetic arcade of unevenly spaced loggia-like columns and a vertically scaled walkway area. The loggia is linked to the gallery space by a stair element providing access to the second-level balcony.

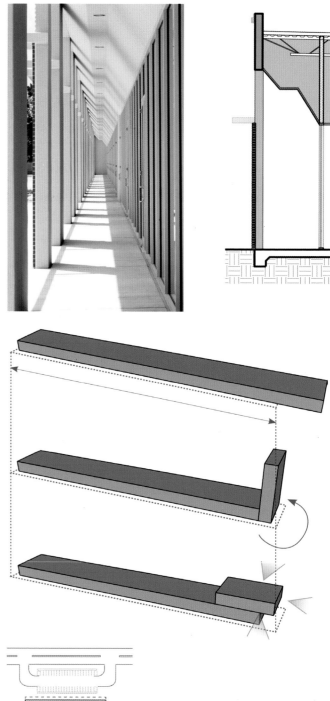

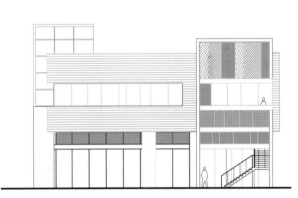

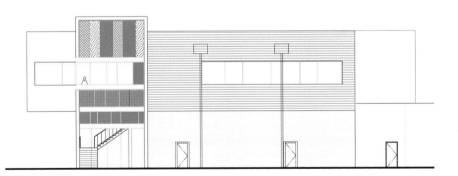

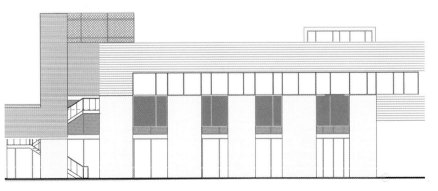

213

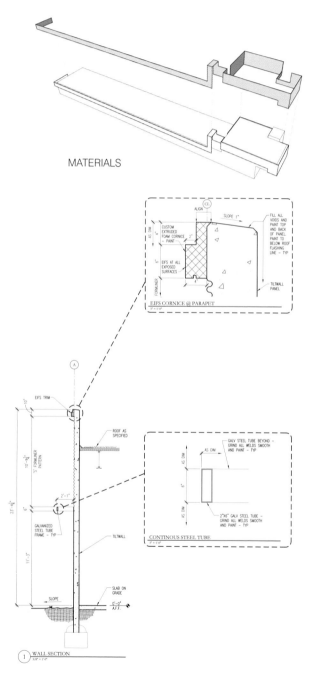

MATERIALS

HEAVY: BRICK
Dark brick selected with gradient variety for the base to balance lighter materials on top.

MEDIUM: FORMLINER
Accent applied to concrete using "S" metal panel profile to give some variety to the flat Tilt Wall panels.

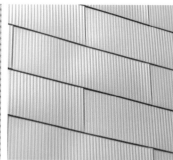

LIGHT: METAL PANELS
Light colored / light material selected for the restaurant box, to make it appear as if held up and locked in placed by heavy brick legs.

The signage band atop the loggia bends vertically to provide an identity opportunity to the joint between the one- and two-story portions of the complex. This gives the building an urban level of activity and articulation.

In a sense, this is a strip retail building with the wrong orientation, making it perpendicular to the street. This however gave the building hierarchy (quite uncommon in strip buildings). The potential for "architecture" at one end and "building" at the other was seemingly retarded by the instruction from the client that it must be built in Tilt Wall, thus no "architecture". In the end it became a demonstration project for just what can be achieved using this method of construction—stacked panels created a void for the curtain-wall box on the west side, site-cast spandrel panels allowed for the floating box cantilever street side, and all was offset by the economics of repetition in the tail potion of the scheme.

Klein

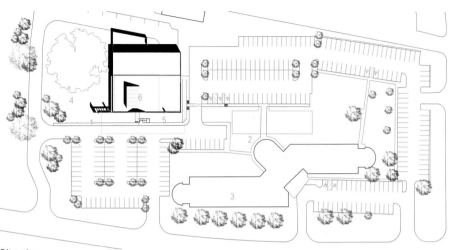

Site plan

1 Footprint of the high school demolished overnight
2 Footprint of the administration building demolished overnight
3 Existing administration building
4 Large oak tree

5 Low roof over administration area
6 Mid-height roof over new network operations center
7 High roof over alumni hall

The background of this problem re-awoke for us of the notion that architecture can be a kind of subsidized self-expression. We are interested in explorations of what is left of Modernism as a valid strategy. Our client reminded us that people often relate to old buildings in a profound way. The high school that had been on this site since the 40s was taken down overnight only to have its ghost met in the morning by many members of this strongly coherent, German-derived community, accompanied by a dirge playing band. They saved what they could (the stone pediment, the bricks from the chimneys, some sports memorabilia) and vowed to fight another day. Over time some parts of the building were lost, but several key fragments survived.

When we were selected to design the new Network Operations Center on this site we met with both the technical client representatives and the alumni association. A deal had been struck: the bond funds were allocated with the stipulation that a space for the alumni to meet was to be included in the facility.

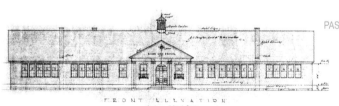

The community outrage over the controversial demolition of the historical high school led us to a solution that was reminiscent of what once was...

...a memory wall [SALVAGED BRICK LIMESTONE PEDIMENT]

Past / present elevations

west elevation

south elevation *memory wall*

east elevation

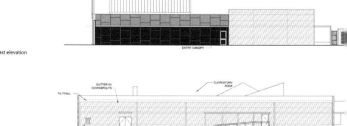

north elevation

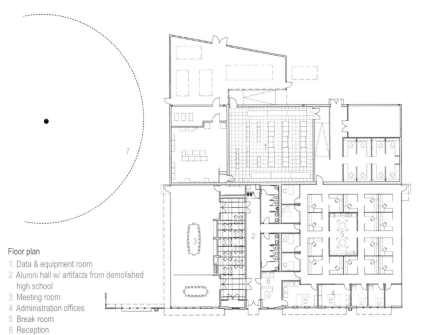

Floor plan

1 Data & equipment room
2 Alumni hall w/ artifacts from demolished
  high school
3 Meeting room
4 Administration offices
5 Break room
6 Reception
7 Large oak

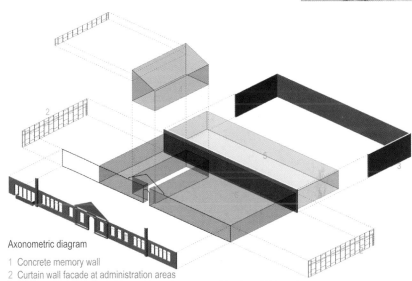

Axonometric diagram

1 Concrete memory wall
2 Curtain wall facade at administration areas
3 Protective concrete shell able to withstand a category 4 hurricane F3 tornado
4 Alumni hall with artefacts from demolished high school
5 Network operations room
6 Administrative functions

Building sections

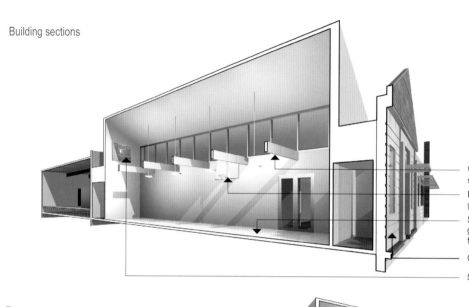

Wood beams from old high school gymnasium

Basketball goal from old gymnasium

Salvaged wood floorboards from gymnasium blended with new flooring

Concrete "memory wall"

Scoreboard from old gymnasium

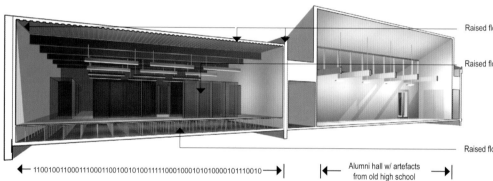

Raised floor distribution area

Raised floor distribution area

Raised floor distribution area

1100100110001110001100100101001111100010001010100001011100101 ←→|

|←→| Alumni hall w/ artefacts from old high school |←→|

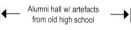

This, of course, was seen as an intrusion into what was meant to be a clinical space for technology and support. The building which houses the IT Department and the Community Alumni Center was to be made of Tilt Wall concrete and metal frame construction, and designed to withstand a Category 4 hurricane and winds from a F3 tornado, allowing the district to maintain communication in the event of such disasters. History be damned.

Our strategy became apparent at once. We utilized a Bakhtin-like concept (he called it "heteroglossia") allowing the subject to speak with multiple voices operating alternately and simultaneously. A memory wall on the exact footings of the original high school is used as a foil against which the Network Operations Center's modern architectural expression coordinates and competes. We operated as if we were concurrently adding on to a building we were designing as new; this facilitated the obvious strategy of addition by contrast. The thinness of the memory wall is intentional. We made it of Tilt Wall concrete 6½ inches thick to encourage it not being perceived as a historical recreation. However, the chimneys are reinterpreted using the original brick, and the limestone pediment is mounted on brackets as if in a museum display. A few of the additional saved elements include the floorboards from the old gymnasium, which were incorporated into the new flooring; the scoreboard, basketball backboard and hoop from the 30s, which were mounted on the wall in the Alumni Center. The window detail in the center was created to mimic the original windows from the old gymnasium.

For us the building doesn't try to recoup the comfortable / familiar qualities of the historic architecture, nor does it attempt to prove that Modernism can be humanistic. Rather it has an unresolved presence that provokes deciphering over interpretation.

Wallis

Wallis State Bank's site is what Mario Gandelsonas termed "ex urban", a site type that we commonly encounter in our work in Houston. At 5 acres and located some 10 miles from the Central Business District, it might be considered suburban, but it is closer to the actual demographic center of Houston, (just a mile south) and is activated by its frontage on the second of now four loops or ring roads (mega highways in actuality) that organize Houston's 624 square mile territory. Thus it is also quite urban and the clients' expectations of what goes on it (program) and how it is organized (form) have more in common with an infill site than a greenfield one.

The site is master-planned for four structures: the bank and administrative facility, a future lease building, a parking garage, and a water-detention facility (further evidence that, despite all, it is an ex urban site). The first phase is a 19,000 square foot floor plate office and bank function. This three-story building also has an executive 4th floor at 5,000 square feet, making a total of 65,000 square feet. It occupies the hard corner and is positioned to have favorable views from the executive level—not to be blocked by the second phase six-story building, (and guarded from the views in by a mesh fin)—with the garage at the "back" of the site buffering the next tier in sites dominated by warehouses.

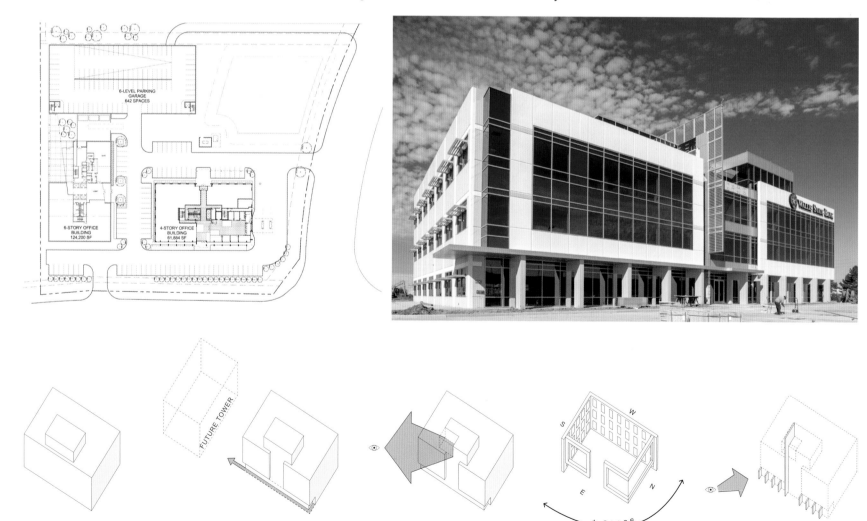

220

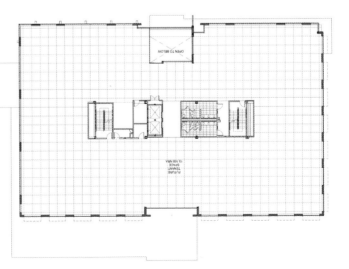

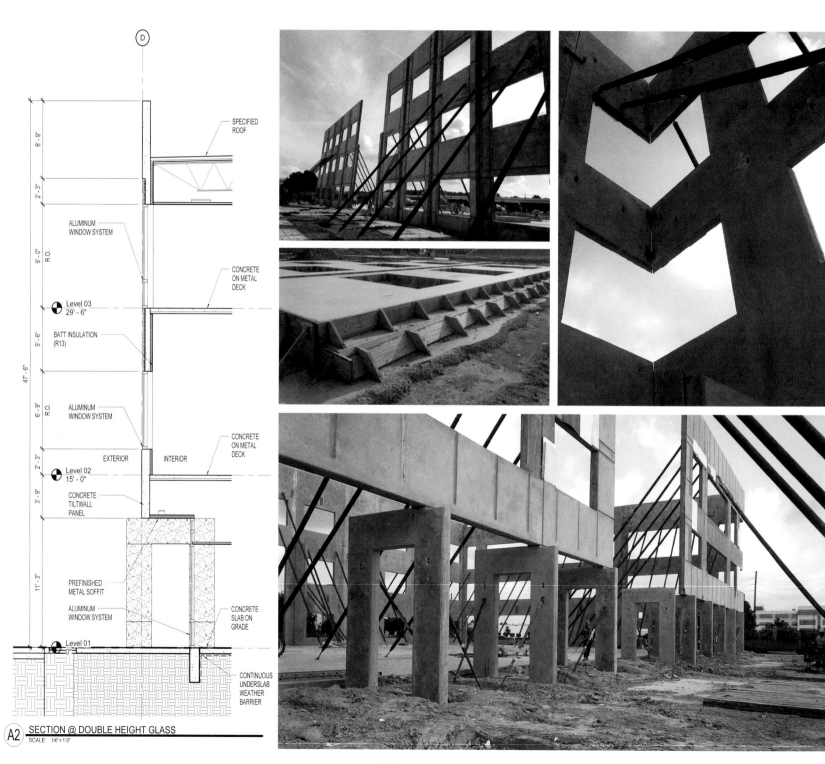

SPECIFIED ROOF

ALUMINUM WINDOW SYSTEM

CONCRETE ON METAL DECK

Level 03
29' - 6"

BATT INSULATION
(R13)

ALUMINUM WINDOW SYSTEM

CONCRETE ON METAL DECK

EXTERIOR   INTERIOR

Level 02
15' - 0"

CONCRETE TILTWALL PANEL

PREFINISHED METAL SOFFIT

ALUMINUM WINDOW SYSTEM

CONCRETE SLAB ON GRADE

Level 01
0"

CONTINUOUS UNDERSLAB WEATHER BARRIER

6' - 9"
2' - 3"
9' - 0" R.O.
5' - 6"
6' - 9" R.O.
2' - 3"
3' - 9"
11' - 3"
47' - 6"

D

A2 SECTION @ DOUBLE HEIGHT GLASS
SCALE: 1/4" = 1'-0"

The skin of the building is organized by both the internal pressures of program organization and a clear solar orientation response. The south and the west elevations are punched openings, while the north and the east combine to form a bent façade which is dominated by glass and glazing. The executive floor is an all-glass box lightly seated atop the base building. The placement of glazing also indexes the site as the façade coincides with the active hard-corner intersection.

The whole skin floats above a loggia space created by a series of Tilt Wall wickets on the front surface with a curtain-wall expression of the entry and lobby area compositionally integrated into the façade plane. The bank drive-through and cover are part of this expression. These elements illustrate what is really the innovation in this project—the dematerialization of the Tilt Wall construction to achieve the desired iconography, as well as the opportunity to capitalize on the economics of a Tilt Wall phase one, as phase two will need to be built conventionally due to site constraints.

The dematerialization approach involves several maneuvers. The south-west skin as mentioned is heavy, punched and has a lower glass-to-wall skin ratio. In contrast, the northeast "floats"—its expression is visually light and it does not touch the ground. The light or tectonic quality of the box comes from combining large-opening-punched panels with cantilevered corner panels and site-cast spandrels. In some cases spandrel glass extends over and hides panel legs giving the appearance of curtain wall. The wickets below are, in some ways, a transformation of Tilt Wall as structure (see the Texas Steel project) to Tilt Wall as infrastructure—that is, they are both load bearing and expressed. Turned vertically to the façade, they provide relief and depth that operates counter to the planarity of Tilt Wall used raw.

CREDITS / PHOTOGRAPHY

All other images © Powers Brown Architecture

BIBLIOGRAPHY

Banham, R., *Theory and Design in the First Machine Age*, MIT Press, Cambridge, 1989.

Benevolo, L., *History of Modern Architecture Vols. 1 and 2*, MIT Press, Cambridge, 1977.

Colquhoun, A., *Modern Architecture*, Oxford University Press, USA, 2002.

Ford, E., *The Details of Modern Architecture*, MIT Press, Cambridge, 1990.

Ford, E., *The Details of Modern Architecture, Volume 2. 1928 to 1988*, MIT Press, Cambridge, 2003.

Frampton, K., *Modern Architecture: A Critical History*, Thames and Hudson, London, 1985.

Frampton, K., *Studies in Tectonic Culture: The Poetics of Construction in Nineteenth and Twentieth Century Architecture*, MIT Press, Cambridge, 1996.

Gebhard, D., *Schindler*, William K. Stout Publishers, San Francisco, 1997.

Ghirado, D. Ed., *Out of Site: A Social Criticism of Architecture*, Bay Press, Seattle, 1991.

Giedion, S., *Space, Time and Architecture*, Harvard University Press, Cambridge, 1954.

Hays, K., Ed., *Architecture Theory since 1968*, MIT Press, New York, 2000.

Jordy, W., *Progressive and Academic Ideals at the Turn of the Twentieth Century*, Oxford University Press, New York, 1972.

Kamerling, B., *Irving J. Gill, Architect*, San Diego Historical Society, San Diego, 1993.

Karlstrom, P. Ed., *On the Edge of America California Modernist Art 1900-1950*, University of California Press, Los Angeles, 1996.

Kwinter, S., *Far From Equilibrium: Essays on Technology and Design Culture*, Actar Press, Barcelona, 2007.

Leach, N., Ed., *Rethinking Architecture: A Reader in Cultural Theory*, Routledge, New York, 1997.

McCoy, E., *Five California Architects*, Hennessey & Ingalls, Santa Monica, 2004.

Nesbitt, K., *Theorizing a New Agenda for Architecture: An Anthology of Architectural Theory 1965-1995*, Princeton Architectural Press, New York, 1996.

Ockman, J., *Architecture Culture 1943-1968: A Documentary Anthology*, Rizzoli, New York, 1993.

Ortega y Gasset, J., *Toward a Philosophy of History*, W.W. Norton & Company, New York, 1941.

Picon, A., "Does Our Technology Make Our Past Irrelevant to Our Future?", *Harvard Design Magazine*, Vol. 31, Fall / Winter, 2009-10.

Rand, M., *Irving J. Gill: Architect, 1870- 1936*, Gibbs Smith Publisher, Salt Lake City, 2006.

Rowe, C., *The Architecture of Good Intentions: Towards a Possible Retrospect*, Academy Editions, London, 1994.

Sauerbruch, M. *Sauerbruch Hutton Archive*, Lars Müller Publishers, Switzerland, 2006.

Smith, R., *Prefab Architecture: A Guide to Modular Design and Construction*, John Wiley & Sons, New Jersey, 2010.

Zaera-Polo, A., "A Scientific Autobiography", *Harvard Design Magazine*, Vol. 21, Fall / Winter, 2004.

INDEX